# CMOS Analog Circuit Design

By YoungHee Kim, PanBong Ha

도서출판 GS인터비전

# Contents

# 1 CHAPTER Semiconductor Properties

The smallest unit of the elements that maintain the chemical properties of a material is called an atom. Each of the 118 already known elements has atoms, the structure of which is different for every other element, or in simple terms each element has its own atomic structure. According to the Bohr's atomic model, an atom consists of nucleus in its center, and electrons circulating the orbit around that nucleus, as shown in Fig. 1, which has negative charge. The nucleus is composed of a proton which is a positive charge, and a non-charged neutron. Elements are electrically neutral because they have the same number of protons and electrons.

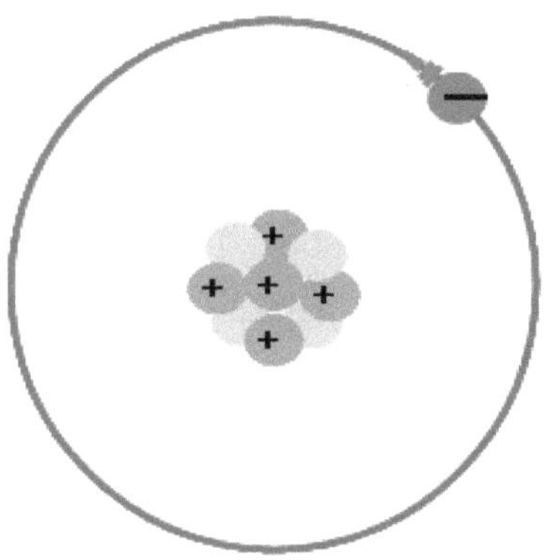

Fig. 1. Bohr's atomic model.

Fig. 2 shows the Bohr's atomic model of silicon. The electrons are orbiting at a certain distance from the nucleus. The electrons circulating in the orbit around the nucleus have far less energy than the electrons circling in farther orbits, and electrons in different orbits of the atom have discontinuous energy levels, and each orbit in the atom corresponds to each shell shown in Fig. 2. Each shell of an atom has a finite number of maximum electrons at the allowed energy level. Shells are numbered 1, 2, 3, and so on, and shell 1 is closest to the atomic nucleus.

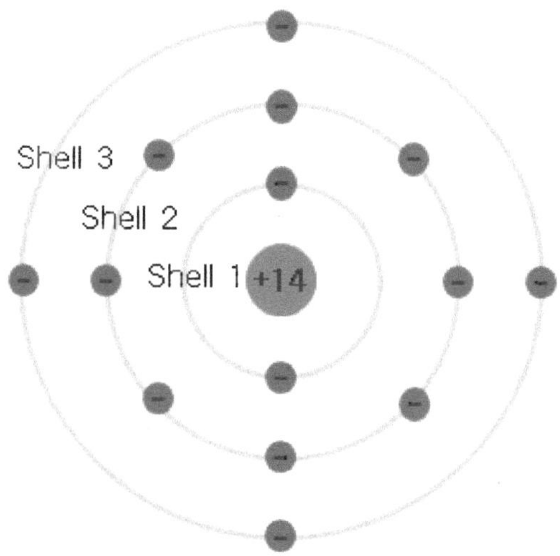


Fig 2. The Bohr's model of silicon atom.

The maximum number of electrons that can exist in the shell of an atom is $Ne = 2n^2$, where n is the shell number. Table 1 shows the maximum number of electrons that can exist in the shell. The electrons in an orbit far from the nucleus have higher energy than the electrons near the nucleus, and the binding force to the nucleus is consequently weak. The electrons with the highest energy level are at the outermost of the atom, and this outermost shell is called the valence shell. The electrons within this valence shell are called valence electrons, and the number of valence electrons determines the properties of the material.

Table 1. The maximum number of electrons that can exist in the shell.

| Shell | Maximum number of electrons |
|---|---|
| 1 | 2 |
| 2 | 8 |
| 3 | 18 |
| 4 | 32 |

The Bohr's model of silicon atom at the ground state of T=0°K is shown in Fig. 2. The number of electrons that can exist in each shell of an atom is as follows: Ne=2 for n=1, Ne=8 for n=2, and Ne=4 for n=3. So, silicon atom is a column IV element which means it is a tetravalent element in the periodic table, since the number of valence electrons are four.

All elements are arranged in the periodic table in atomic number order. The atomic number is equal to the number of protons in the nucleus. Table 2 shows the periodic table of the elements. The atomic number of hydrogen is 1 and the atomic number of helium is 2. And the atomic number of boron, carbon, nitrogen, aluminum, silicon and phosphorus is 5, 6, 7, 13, 14 and 15, respectively.

Table 2. Periodic table of elements.

| I | II | III | IV | V | VI | VII | VIII |
|---|---|---|---|---|---|---|---|
| H | | | | | | | He |
| Li | Be | B | C | N | O | F | Ne |
| Na | Mg | Al | Si | P | S | Cl | Ar |
| K | Ca | | | | | | |

A semiconductor refers to a material having a property of a conductor that conducts electricity well or a property of an non-conductor that does not conduct electricity, according to electrical conditions such as voltage and current. Semiconductor materials are column IV of the periodic table and the materials next to it. Silicon and germanium, which belong to column IV, are called elemental semiconductors and composed of single species of atoms. Table 3 shows the location of the semiconductor material in the periodic table of the elements.

Table 3. Periodic table of elements of semiconductor materials.

| II | III | IV | V | VI |
|---|---|---|---|---|
| | B | C | N | |
| | Al | Si | P | S |
| Zn | Ga | Ge | As | Se |
| Cd | In | | Sb | Te |

In addition to the elemental semiconductors, compounds of atoms belonging to column III and column V of the periodic table, or certain combinations of some atoms belonging to column II and column VI, or compounds of column IV atoms, make up the compound semiconductors. The binary compound semiconductors shown in Table 4 are III-V compound semiconductors, II-VI compound semiconductors, and IV compound semiconductors. In addition to two-element compound semiconductors, there are three-element compound semiconductors such as GaAsP, and quaternary compound semiconductors such as InGaAsP.

Table 4. Two-element compound semiconductors.

| Ⅳ Compounds | Ⅲ-Ⅴ Compounds | Ⅱ-Ⅵ Compounds |
|---|---|---|
| SiC | AlP | ZnS |
| SiGe | AlAs | ZnSe |
| | AlSb | ZnTe |
| | GaN | Cds |
| | GaP | CdSe |
| | GaAs | CdTe |

A single independently-present silicon atom has four outermost electrons and requires four more electrons to form a complete orbit. To form silicon crystals, each atom shares one outermost electron with each of the four adjacent silicon atoms to form a covalent bond. Fig. 3 shows the two-dimensional covalent bond structure of silicon.

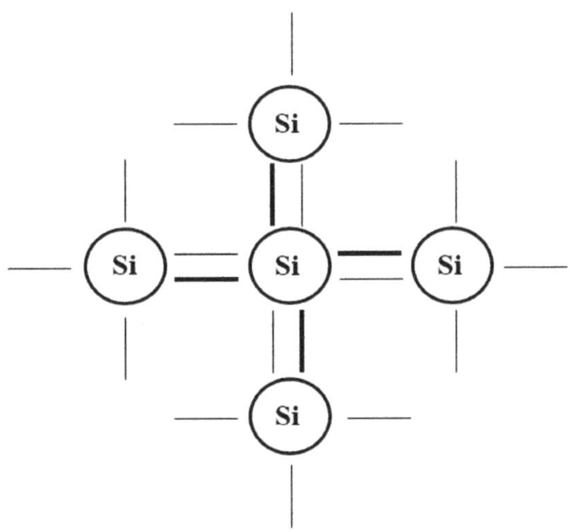


Fig. 3. Two-dimensional covalent bond structure of silicon.

According to the Bohr's model, electrons are in any stable circular orbit around the nucleus for single independently-present atom. The electrons can then be transitted to higher or lower energy orbits, thereby acquiring or losing energy, such as the difference in energy level corresponding to the trajectory before and after the transition.

When a solid is formed by approaching atoms that are separated from each other, various interactions occur between adjacent atoms. The attractive force and the repulsive force between the atoms will equilibrate at the appropriate atomic distance of the crystal. As the distance between the atoms approaches the interatomic gap at the equilibrium of the silicon crystal, it is divided into two bands which are separated by the bandgap energy Eg as shown in Fig. 4.

The energy band in which the valence electrons are present is called the valence band, and the band above the valence band is called the conduction band. Between the valence band and the conduction band, there is the forbidden band that can occupy no energy level of electrons. The height of the forbidden band is called the bandgap energy, and the bandgap energy is the minimum energy required for the valence electron to break the covalent bond and move to the conduction band. For example, the bandgap energy of silicon is about 1.1eV.

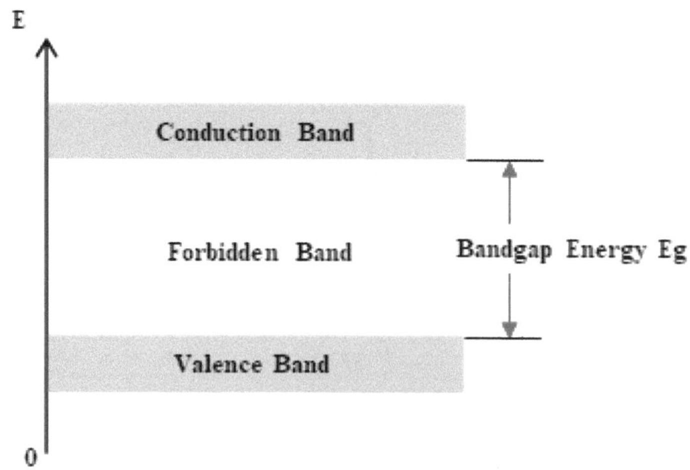


Fig. 4. Energy band diagram.

At the ground state of T=0˚K, the valence band is completely filled with electrons, as shown in Fig. 5, and the conduction band is emptied. Since there is no empty state in which electrons can travel, there is no transfer of charges within the valence band. In addition, there is no electron in the conduction band, so there is no transfer of charges as well.

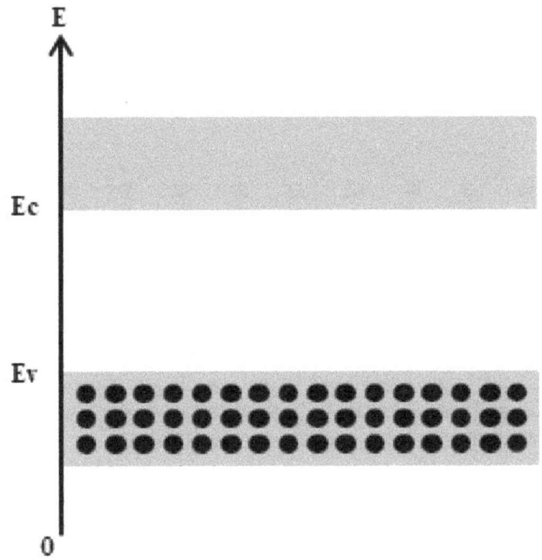


Fig. 5. Energy band diagram at T=0˚K.

Charged particles moving in a semiconductor are called charge carriers, and there are two types of charge carriers such as electrons and holes. The electron refers to a particle with negative charge $q = -1.6 \times 10^{-19}$C, and the hole refers to a positive charge with the same charge but opposite sign to the aforementioned negative charge. As the temperature of the semiconductor rises from 0°K, thermal energy larger than the bandgap energy is applied.

When a part of the electrons are excited and move to the conduction band through the energy bandgap, the same number of holes are generated in the valence band. If electrons in the conduction band and holes in the valence band are generated by thermal excitation, they are called EHP (Electron-Hole Pair) generation. Fig. 6 shows the EHP generation due to thermal energy in semiconductors. The electrons in the conduction band and the holes in the valence band move independently of each other. The black colored circles shown in the conduction band of Fig. 6 are electrons and the vacant white circles shown in the valence band are holes.

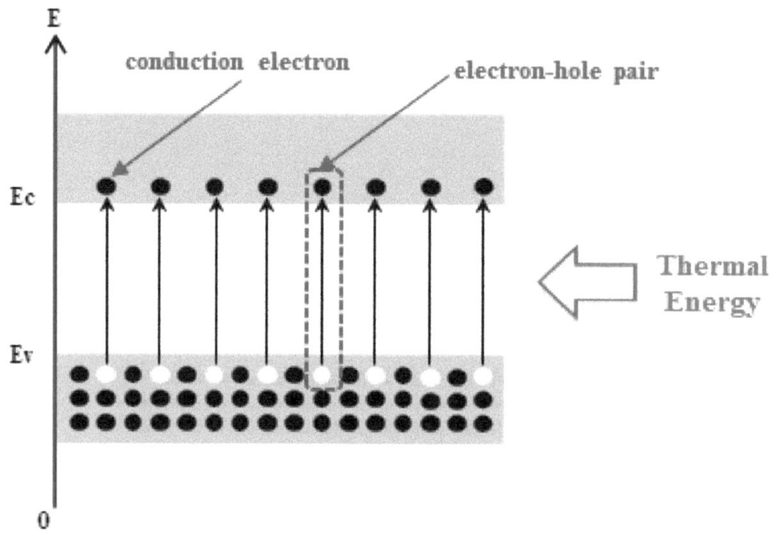


Fig. 6. EHP generation due to thermal energy in semiconductors.

Semiconductors are classified into intrinsic semiconductors and extrinsic semiconductors depending on whether impurities are doped or not. A complete semiconductor crystal without any impurities or lattice defects is called an intrinsic semiconductor. The charge carriers carrying the charges in the intrinsic semiconductor are thermally generated electrons and holes. Since electrons and holes are generated in pairs, the electron concentration n of the conduction band (number of electrons per ㎤) is the same as the hole concentration p (number of holes per ㎤) in the valence band and is expressed as $n_i$, which is the intrinsic concentration. Obviously, to maintain a steady-state carrier concentration, there must be recombination of electron-hole pairs at the same rate as the electron-hole pairs are generated. At room temperature of 300°K, $n_i$ in intrinsic silicon is only $1.5\times10^{10}$ EHPs/㎤, much less than the silicon atom density of $5\times10^{22}$ atoms/㎤. Therefore, intrinsic semiconductors have very small current and are close to insulator properties.

$$n = p = n_i \tag{1}$$

Generally, to make semiconductor devices such as diodes and transistors, extrinsic semiconductors is doped with an appropriate amount of impurities of column III or impurities of column V in an intrinsic semiconductor. The electrical properties of extrinsic semiconductors are determined by the type of impurities and the doping concentration. An extrinsic semiconductor is classified into an n-type semiconductor and a p-type semiconductor depending on the type of doped impurities.

When column V impurities (P, As, and Sb) are doped into the intrinsic silicon to increase the number of electrons in the conduction band, a column V impurity atom replace the silicon atom in the crystal lattice as shown in Fig. 7.

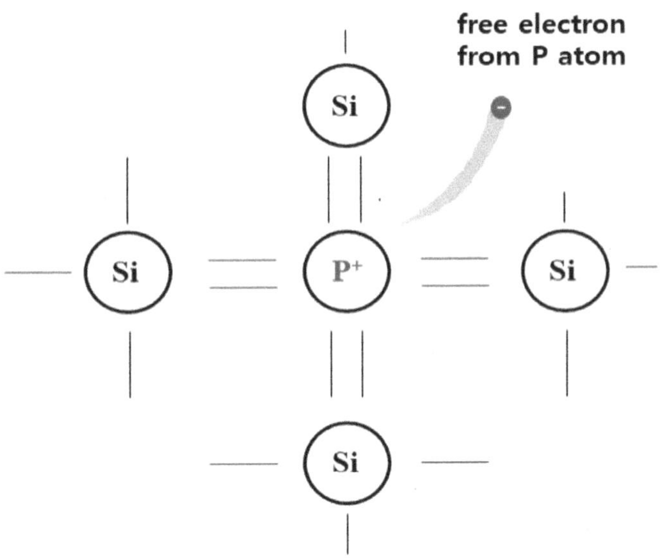


Fig. 7. An n-type semiconductor crystal structure of silicon.

At T = 0°K, four of the five valence electrons from a column V impurity atom share a covalent bond with the neighboring four silicon atoms, and the remaining electron is located on a donor level near the conduction band. This donor level is filled with electrons at 0°K as shown in Fig. 8(a), and even if extremely small thermal energy of about 50 to 100°K is applied, they are donated to the conduction band as shown in Fig. 8(b). Thus, the impurity level of the column V is called the donor level, and the column V impurity is called the donor impurity. Once the donor has made the free electrons, the donor impurities, which are column V impurities, are ionized positively and participate in the covalent bond.

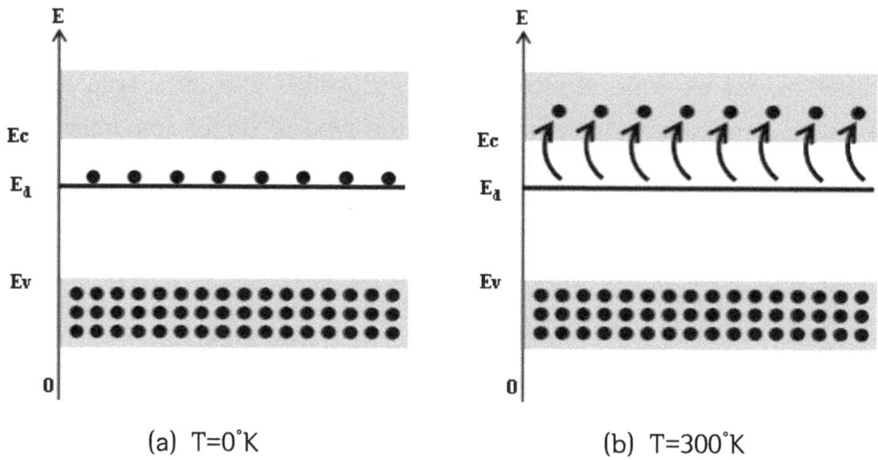


Fig. 8. Energy band diagram of n-type semiconductor.

In the n-type semiconductor doped with the donor impurities, the electron concentration $n_0$ in the conduction band at room temperature is much larger than the intrinsic concentration $n_i$ or the hole concentration $p_0$ in the valence band, and the majority carriers in the silicon semiconductor are electrons. The minority carriers are holes. The intrinsic semiconductor doped with the donor impurities is an n-type semiconductor because the electrons with negative charges are majority carriers.

When column III impurities (B, Ga, and In) are added to the intrinsic silicon to increase the number of holes in the valence band, column III impurities atom enter the silicon lattice in the crystal lattice as shown in Fig. 9. A column III impurity atoms make an acceptor level near the valence band in silicon.

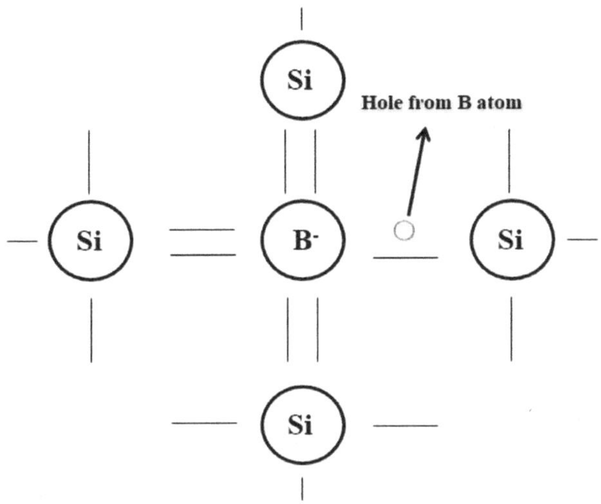


Fig. 9. p-type semiconductor crystal structure of silicon.

At low temperatures, if sufficient thermal energy is provided to a column III impurity atoms, electrons can be excited from the valence band to the acceptor level and then holes in the valence band are left, as shown in Fig. 10(b). The acceptor level accepts electrons from the valence band, and the column III impurities are the acceptor impurities.

In the p-type semiconductor doped with the acceptor impurities, the hole concentration $p_0$ in the valence band at room temperature is much larger than the intrinsic concentration $n_i$ or the electron concentration $n_0$ in the conduction band. In the silicon semiconductor, the majority carriers are holes and the minority carriers are electrons. The intrinsic semiconductor doped with the acceptor impurities is a p-type semiconductor because the holes with positive charges are majority carriers.

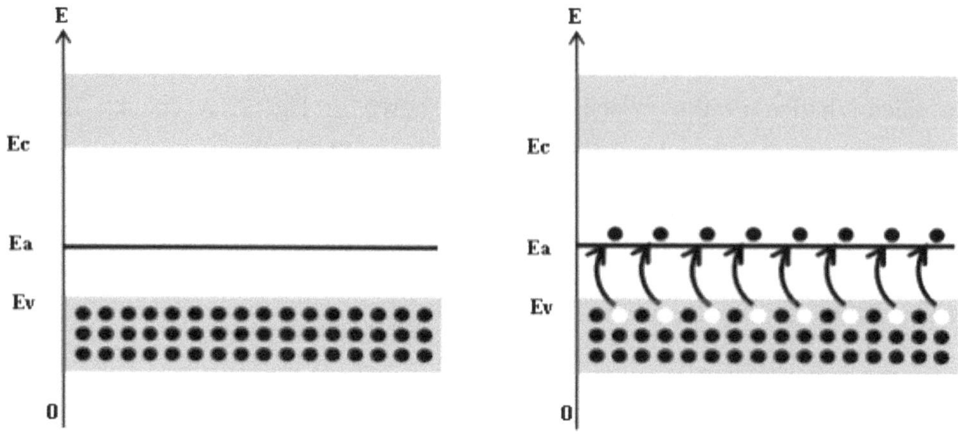


(a) T=0˚K (b) T=300˚K

Fig. 10. Energy band diagram of p-type semiconductor.

In general, impurities which make n-type and p-type semiconductors exist in the semiconductor at the same time, and the semiconductor type is determined according to which dopants are doped much more. When the donor impurity concentration $N_d$ is much larger than the acceptor impurity concentration $N_a$, the intrinsic semiconductor doped with the donor impurities and acceptor impurities is an n-type semiconductor having an electron concentration of $N_d - N_a$, and when $N_a \gg N_d$, it is a p-type semiconductor. When $N_d = N_a$, it is a completely compensated semiconductor. In this case, since the concentration of electrons and holes due to doping impurities from the outside is the same, it becomes an intrinsic semiconductor. Fig. 11 shows the semiconductor types according to the doping concentration. In Fig. 11(a), the part marked with $N_d$ is a case where the donor impurity concentration $N_d$ is doped larger than the acceptor impurity concentration $N_a$. The region doped with donor impurities is an n-type semiconductor having an electron concentration corresponding to $N_d - N_a$. On the other hand, in Fig. 11(b), the part marked with $N_a$ is a case where the acceptor impurity concentration $N_a$ is doped larger than the donor impurity concentration $N_d$. The region doped with acceptor impurities is an p-type semiconductor having an hole concentration corresponding to $N_a - N_d$. Lastly, Fig. 11(c) shows the case where the donor impurity concentration $N_d$ and the acceptor impurity concentration $N_a$ are injected at the same concentration in intrinsic silicon, and it has intrinsic silicon properties even when impurities are injected.

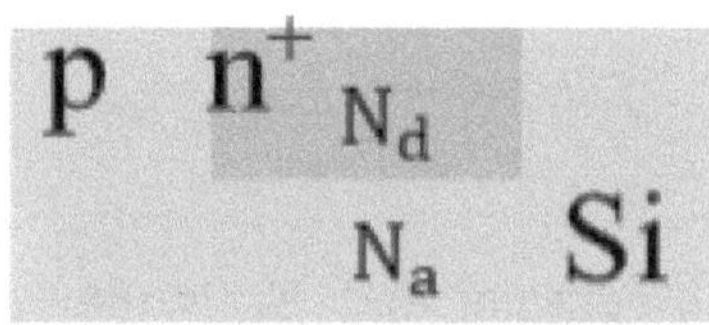


(a) in case of $N_d$ >> $N_a$

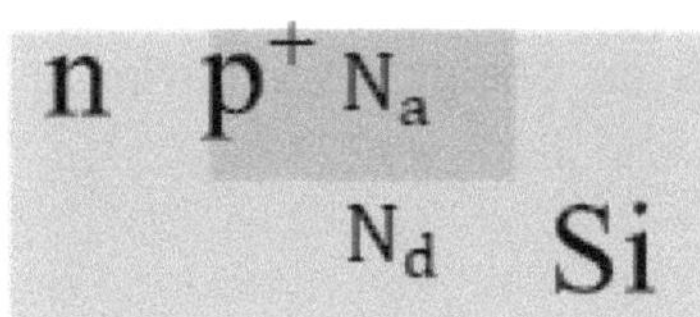


(b) in case of $N_a$ >> $N_d$

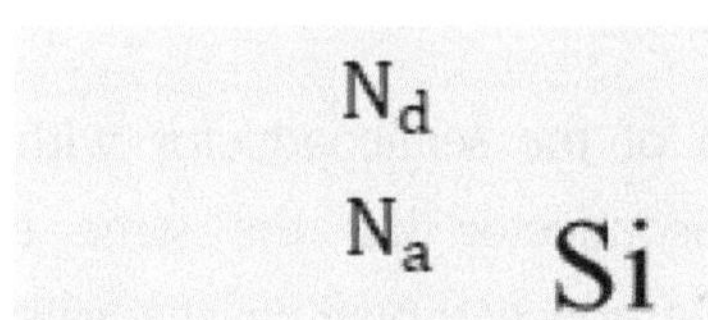


(c) in case of $N_a$ = $N_d$

Fig. 11. Classification of the semiconductor types according to the doping concentration.

In calculating the electrical properties of semiconductors and analyzing the behavior of semiconductor devices, it is necessary to know the number of charge carriers per ㎤ in the material. The electrons in the solids are in accordance with the Fermi-Dirac statistics, and the results from the discussion of this statistics show that the electron distribution in the energy level range in the thermal equilibrium state is as shown in Eq. (2).

$$f(E) = \frac{1}{1+e^{(E-E_F)/kT}} \tag{2}$$

where k is the Boltzmann constant and $E_F$ is the Fermi level.

The Fermi-Dirac distribution function f(E) in Eq. (2) is the probability that the electron occupies the energy E at the absolute temperature T. As shown in Eq. (3), the Fermi level $E_F$ represents an important amount for analyzing the operation of the semiconductor which is an energy level having a probability of being occupied by electrons of 1/2.

$$f(E_F) = \frac{1}{1+e^{(E_F-E_F)/kT}} = \frac{1}{1+e^0} = \frac{1}{1+1} = \frac{1}{2} \tag{3}$$

The energy band diagram of the semiconductor with the Fermi level is shown in Fig. 12. The intrinsic semiconductor has the same electron concentration in the conduction band as the hole concentration in the valence band, so the Fermi level $E_F$ is located at the center of the bandgap as shown in Fig. 12(a). In the n-type semiconductor, the electron concentration in the conduction band is larger than the hole concentration in the valence band, so the Fermi level $E_F$ is located near the $E_c$ level as shown in Fig. 12(b). On the other hand, since the hole concentration of the p-type semiconductor is larger than the electron concentration of the conduction band, the Fermi level $E_F$ is located near the $E_v$ level as shown in Fig. 12(c).

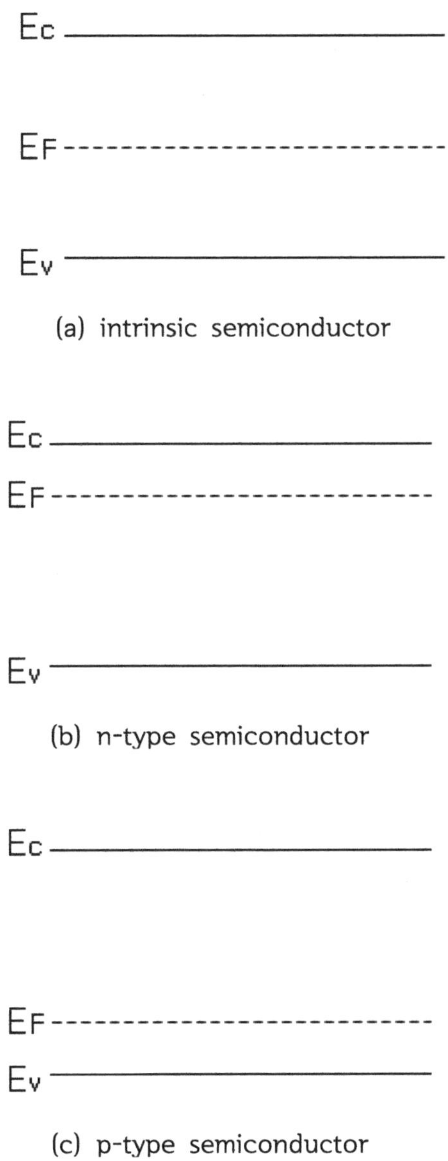


Fig. 12. Semiconductor energy band diagram considering Fermi level.

In the thermal equilibrium state, the electron concentration in the conduction band and the hole concentration in the valence band can be expressed by Eqs. (4) and (5), where $E_i$ is the Fermi level in the intrinsic semiconductor.

$$n_0 = n_i e^{(E_F - E_i)/kT} \tag{4}$$

$$p_0 = n_i e^{(E_i - E_F)/kT} \tag{5}$$

**Example 1**

*What is the concentrations of $n_0$ and $p_0$ respectively in n-type semiconductor doped with donor impurities of $N_d = 10^{16}$ donors/㎤ at T=300 ㎝K?*

For an n-type semiconductor with a donor impurity concentration $N_d=10^{16}$ donors/㎤ in a silicon sample, the electron concentration $n_0$ in the conduction band at 300°K can be approximated as shown in Eq. (6).

$$n_0 \cong N_d = 10^{16}\ electrons/cm^3 \tag{6}$$

$p_0$ is calculated from a relation equation of $n_0 p_0 = n_i^2$ as Eq. (7).

$$p_0 = \frac{n_i^2}{n_0} \cong \frac{n_i^2}{N_d} = 2.25 \times 10^4\ holes/cm^3 \tag{7}$$

$E_F - E_i$ is calculated as Eq. (8), and the energy band diagram using it is shown in Fig. 13.

$$E_F - E_i = kT \cdot \ln\left(\frac{n_0}{n_i}\right) = 0.0259 \cdot \ln\left(\frac{10^{16}}{1.5\times 10^{10}}\right) = 0.347\ eV \quad (8)$$

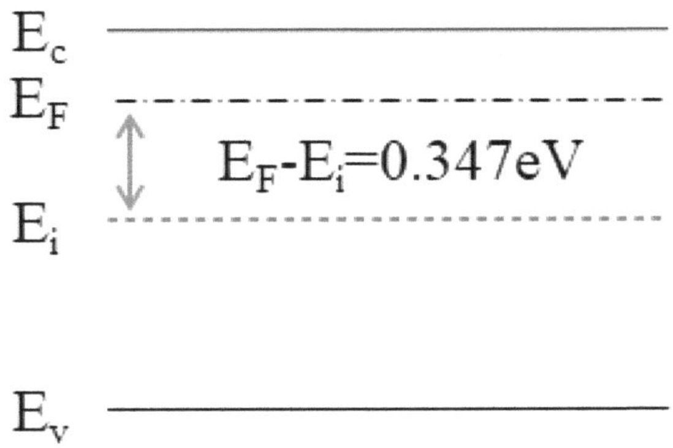


Fig. 13. Energy band diagram for the case of n-type semiconductor with donor impurity concentration $N_d = 10^{16}$ donors/㎤.

## Exercises

1. Answer the following questions.

(a) Write the electron configuration for Si in the ground state.

(b) What is the $n_i$ of Si at room temperature?

(c) Complete the table below.

| | majority carrier | minority carrier |
|---|---|---|
| n-type | | |
| p-type | | |

(d) What is the bandgap energy of Si?

(e) What are $p_0$ and $n_0$ in case of Si doped with impurity of $N_a = 10^{16}$ acceptors/㎤ at T=300°K?

(f) What is $E_F - E_i$ in case of Si doped with impurities of $N_d = 10^{17}$donors/㎤ at T=300°K?

2. Answer the following questions.

(a) What is the energy level where the probability of finding electrons is 1/2?

(b) What are $p_0$ and $n_0$ in case of n-type Si doped with impurities of $N_d = 10^{15}$ donors/㎤ at T=300°K?

(c) What is $E_i - E_F$ in case of p-type Si doped with impurities of $N_a = 10^{15}$ acceptors/㎤ at T=300°K?

(d) Write Fermi-Dirac distribution function f(E).

(e) What is the bonding strength of silicon solid?

(f) What is the thermal voltage at T=300°K?

(g) What column element is P in the periodic table?

(h) What is the current by the electric field?

(i) What is the current proportional to the slope of the carrier concentration?

(j) Write the electron configuration table of C at T=0°K.

(k) What column element is B in the periodic table?

(l) What is the maximum number of electrons that can fit in a shell with n equal to 2?

(m) What is the probability of finding electrons at the Fermi level?

(n) What is the absolute temperature T in the ground state (in the unit of °K)?

(o) What is kT at T=300°K?

3. Assume that a Si sample is doped with $5\times10^{15}$donors/㎤ at T=300°K.

(a) Calculate $n_0$ and $p_0$ in the thermal equilibrium state.

(b) Calculate $E_F - E_i$.

(c) Draw the energy band diagram.

4. Answer the following questions for Si semiconductor doped with $N_a$ = $5\times10^{15}$ acceptors/㎤.

(a) What are $n_0$ and $p_0$ at T=0°K?

(b) What are $n_0$ and $p_0$ at T=300°K?

(c) What is $E_i - E_F$ at T=300°K?

(d) Draw the energy band diagram at T=300°K.

## Chap. 1 References

[1] Sanghee Son et al., Electronic Circuits, First Book, 2015.

[2] Kyungwook Shin, Core Electronic Circuit, Hanbit Academy, 2012.

[3] Gyedal Kwak et al., Solid-State Electronics, Pearson, 2009.

[4] Hojeong Hwang, Semiconductor Engineering, Saengneung, 2014.

CHAPTER 2

# pn Junction Diodes

## 2.1 Equilibrium State

When the p-type semiconductor and the n-type semiconductor are separated as shown in Fig. 1, the majority carriers in the p-type semiconductor are holes and the minority carriers are electrons. In the n-type semiconductor, the majority carriers are electrons and the minority carriers are holes.

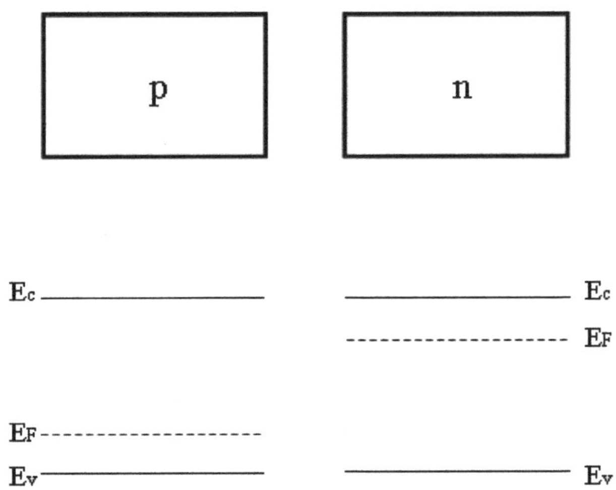


Fig. 1. Energy band diagram for p-type and n-type semiconductors before bonding.

Fig. 2 shows a conceptual representation of the phenomenon occurring over time in the process of forming a pn junction. At the moment of t=0 when the p-type semiconductor is bonded with the n-type semiconductor as shown in Fig. 2(a), there are many holes in the p region and many electrons in the n region. From the moment the pn junction is formed, the holes recombine with the electrons near the junction of the n region, while diffusing from the p region of high concentration to the n region of low concentration. On the other hand, the electrons diffuse from the n region to the p region and recombine with the holes near the junction of the p region.

As shown in Fig. 2(b), at time $t = t_1$, the holes generated from the acceptor impurities in the p region diffuse into the n region, and the acceptor impurities in the p region near the junction form a negative ion layer by losing their holes, while the donor impurities in the n region near the junction form a positive ion layer while the electrons generated from the donor impurities in the n region are diffused into the p region.

The positive and negative ion layers near the pn junction form a depletion region. The term depletion is used in the sense that charge carriers, such as electrons and holes, are diffused across the pn junction and that the charge carriers are depleted in the depletion region near the pn junction, and the depletion regions are formed in an instant. Diffusion currents move large amounts of charges from one direction to the other, which eventually converge to zero. This is because the pn junction terminal is open and can not flow indefinitely. As shown in Fig. 2(c), the depletion width does not expand more at the equilibrium state of t=∞.

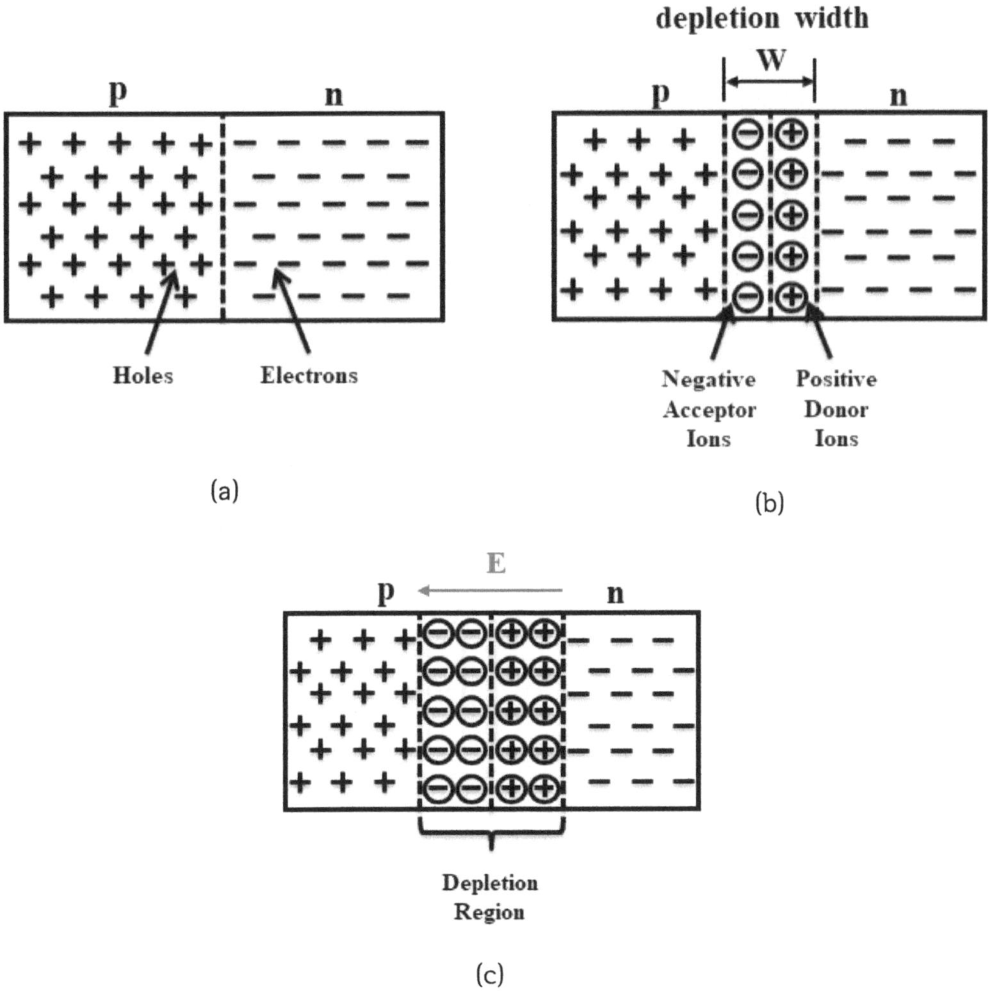


Fig. 2. The formation process of depletion region in pn junction: (a) moment of bonding ($t$=0), (b) $t_1$ after bonding ($t = t_1$), and (c) equilibrium state ($t$=∞).

As shown in Fig. 3, an electric field is formed from the positive ions directed to the negative ions in the depletion region, and this electric field serves to stop the injection of holes into the n region and the injection of electrons into the p region. Thus, the electric field formed in the depletion region acts as a potential barrier that prevents the charge carriers from injecting to other regions. Therefore, after a certain time has elapsed from the time $t = 0$ at which the pn junction is formed, the charge carriers are in an equilibrium state in which they can no longer inject to other regions.

As shown in Fig. 3, the voltage difference $V_0$ generated between the depletion regions is referred to as a contact voltage. The contact voltage $V_0$ across the depletion width W is related to the acceptor impurity concentration of the p region and the donor impurity concentration of the n region as shown in Eq. (1).

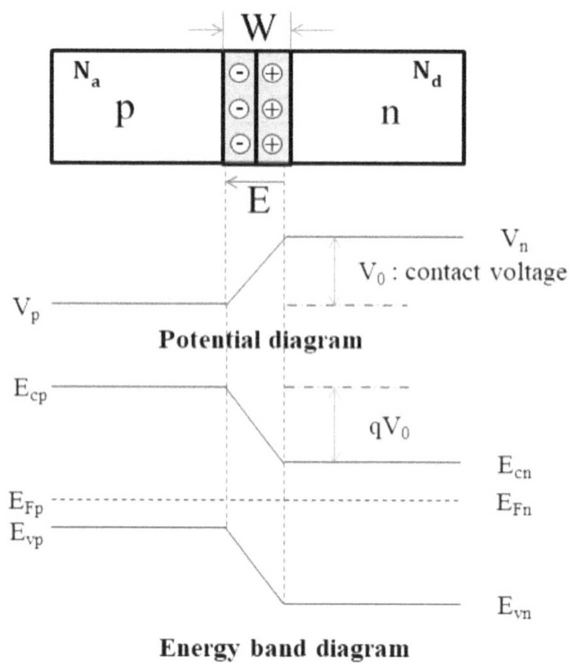


Fig. 3. Depletion charges in the depletion region W, resulting electric field E; and electrostatic potential and energy band diagram.

$$V_0 = V_T \ln \frac{N_a N_d}{n_i^2} \tag{1}$$

In the potential diagram in Fig. 3, the p and n regions outside the depletion region are electrically neutral so that the potential $V_p$ and potential $V_n$ are constant when the electric field is zero from Gauss's law. In the depletion region, $V_n$ is higher than $V_p$ by $V_0$. In addition, the energy band diagram at equilibrium in the pn junction diode first matches the Fermi level, then the energy band diagram in the p region brings the enfergy band diagram of the p-type semiconductor shown in Fig. 1, and the energy band in the n region does the energy band diagram of the n-type semiconductor shown in Fig. 1, by directly importing the energy band diagrams of both semiconductors. In the depletion region, the conduction energy level $E_c$ of the p-type semiconductor is connected to the $E_c$ of the n-type semiconductor, and the valence energy level $E_v$ of the p-type semiconductor is connected to the $E_v$ of the n-type semiconductor. The energy band diagram of the electron shown in Fig. 2 shows that the $E_{cp}$ of the p-type semiconductor is q$V_0$ higher than that of the n-type semiconductor. Therefore, the energy band diagram in the equilibrium state has many electrons in the conduction band of the n-type semiconductor, but the electrons are not injected from the n-type semiconductor to the p-type semiconductor because $E_{cn}$ is lower than $E_{cp}$. Also, there are many holes in valence band of p-type semiconductor, but since $E_{vp}$ is higher than $E_{vn}$, holes cannot be injected from p-type semiconductor to n-type semiconductor.

In Fig. 4, the pn junction is doped with $N_a \rangle N_d$, where $x_{p0}$ is the depletion width in the p region and $x_{n0}$ is the depletion width in the n region. The depletion width W is the sum of $x_{p0}$ and $x_{n0}$. W, $x_{p0}$, and $x_{n0}$ are expressed by Eqs. (2), (3), and (4), respectively. As can be seen from Eqs. (3) and (4), the depletion region is mainly formed in the lightly doped region.

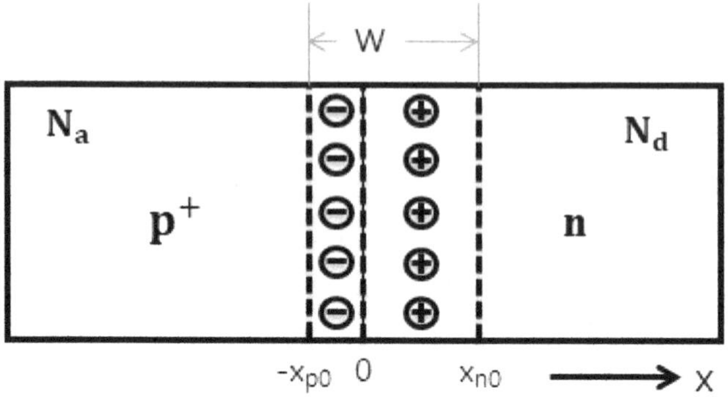


Fig. 4. Process cross-sectional view of a pn junction diode doped with $N_a > N_d$.

$$W = \sqrt{\frac{2\epsilon_0\epsilon_{Si}V_0}{q}\left(\frac{1}{N_a}+\frac{1}{N_d}\right)} \tag{2}$$

$$x_{p0} = \frac{N_d}{N_a+N_d}W \tag{3}$$

$$x_{n0} = \frac{N_a}{N_a+N_d}W \tag{4}$$

As shown in Fig. 5, abrupt pn junction diode refers to a diode that suddenly changes from p-type acceptor impurity concentration $N_a$ to n-type donor impurity concentration $N_d$ at the pn junction. When the voltage across the pn junction diode is changed, the capacitance exists because the depletion charges in the depletion region change, and the capacitance is called the junction capacitance. Junction capacitance shown in eq. (5) is proportional to junction area A and inversely proportional to the depletion width W where the proportionality constant is $\epsilon_0\epsilon_{Si}$. Here, $\epsilon_0$ is the dielectric constant in vacuum and $\epsilon_{Si}$ is the dielectric constant of silicon.

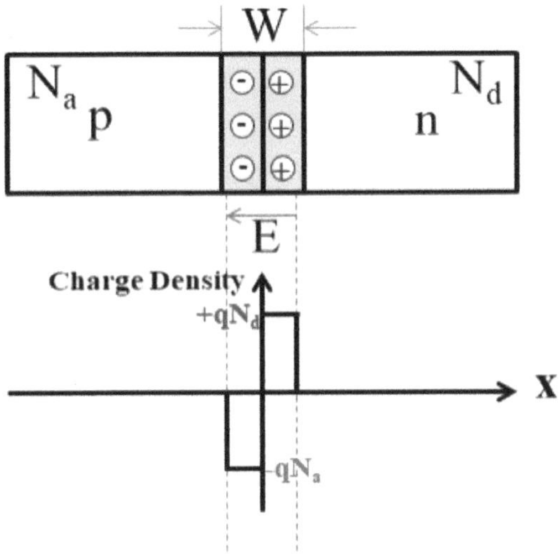


Fig. 5. Charge density curve at abrupt pn junction diode.

$$C_j = \epsilon_0 \epsilon_{Si} \frac{A}{W} \tag{5}$$

Example 2-1

*The impurity concentrations of the p region and the n region at the silicon pn junction are doped with* $N_a = 10^{17}$ *acceptors/㎤ and* $N_d = 5 \times 10^{15}$ *donors/㎤ respectively. Answer the following questions.*

(a) Calculate $E_{ip} - E_{Fp}$, $E_{Fn} - E_{in}$, and $V_0$.

$E_{ip} - E_{Fp}$ and $E_{Fn} - E_{in}$ are calculated at 300°K as shown in Eqs. (6) and (7), respectively. The contact potential $V_0$ can be calculated as shown in Eq. (8).

$$E_{ip} - E_{Fp} = kT \times \ln\frac{p_0}{n_i} = 0.0259 \times \ln\frac{10^{17}}{1.5 \times 10^{10}} = 0.407e\,V \tag{6}$$

$$E_{Fn} - E_{in} = kT \times \ln\frac{n_0}{n_i} = 0.0259 \times \ln\frac{5 \times 10^{15}}{1.5 \times 10^{10}} = 0.329e\,V \tag{7}$$

$$V_0 = \frac{kT}{q} ln\frac{N_a N_d}{n_i^2} = 0.0259 \times \ln\frac{10^{17} \times (5 \times 10^{15})}{(1.5 \times 10^{10})^2} \tag{8}$$
$$= 0.736\,V$$

(b) Draw the energy band diagram of the pn junction diode.

Fig. 6 shows the energy band diagram of the pn junction diode using $E_{ip} - E_{Fp}$, $E_{Fn} - E_{in}$, and $V_0$ obtained in solution (a). Fig. 6 shows an energy band diagram of a pn junction diode with $N_a = 10^{17}$ acceptors/㎤ and $N_d = 5 \times 10^{15}$ donors/㎤ at 300°K. The energy values of $E_{ip} - E_{Fp}$, $E_{Fn} - E_{in}$, and $E_{cp} - E_{cn}$ are shown in the energy band diagram, and $E_{cp}$ is 0.736 eV higher than $E_{cn}$ in the energy band diagram. This means that $V_n$ is 0.736 V higher than $V_p$.

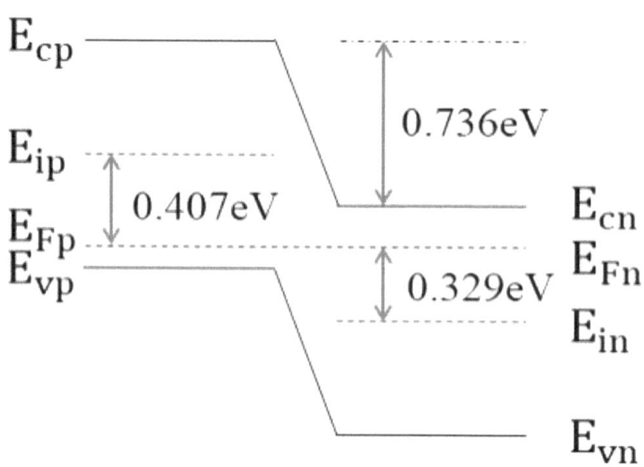


Fig. 6. Energy band diagram at pn junction.

(c) Calculate the depletion width W, $x_{p0}$, and $x_{n0}$.

$W$, $x_{p0}$ and $x_{n0}$ in the pn junction diode can be calculated by using Eqs. (2), (3) and (4). The process of the calculated $W$, $x_{p0}$ and $x_{n0}$ is shown in Eqs. (9), (10) and (11), respectively at 300°K when the silicon pn junction is doped with $N_a = 10^{17}$ acceptors/㎤ and $N_d = 5\times10^{15}$ donors/㎤.

$$\begin{aligned} W &= \sqrt{\frac{2\epsilon_0\epsilon_{Si}V_0}{q}\left(\frac{1}{N_a}+\frac{1}{N_d}\right)} \\ &= \sqrt{\frac{2\times11.8\times8.854\times10^{-14}\times0.736}{1.6\times10^{-19}}\left(\frac{1}{10^{17}}+\frac{1}{5\times10^{15}}\right)} \\ &= 0.449\mu m \end{aligned} \tag{9}$$

$$\begin{aligned} x_{p0} &= \frac{N_d}{N_a+N_d}W \\ &= \frac{5\times10^{15}}{10^{17}+5\times10^{15}}\times0.449\ \mu m \\ &= 0.0214\ \mu m \end{aligned} \tag{10}$$

$$\begin{aligned} x_{n0} &= \frac{N_a}{N_a+N_d}W \\ &= \frac{10^{17}}{10^{17}+5\times10^{15}}\times0.449\ \mu m \\ &= 0.428\ \mu m \end{aligned} \tag{11}$$

**Example 2-2**

*Answer the following questions for a silicon pn junction diode with doping concentrations of* $N_d = 10^{16}$ *donors/㎤ and* $N_a = 10^{16}$ *acceptors/㎤ at T=300°K.*

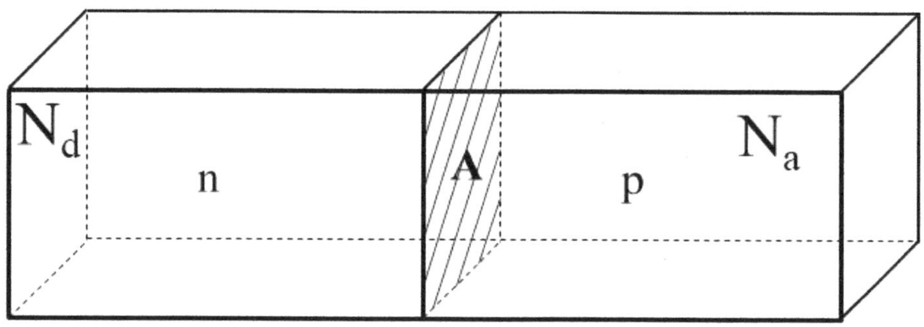


(a) pn junction diode with junction area A

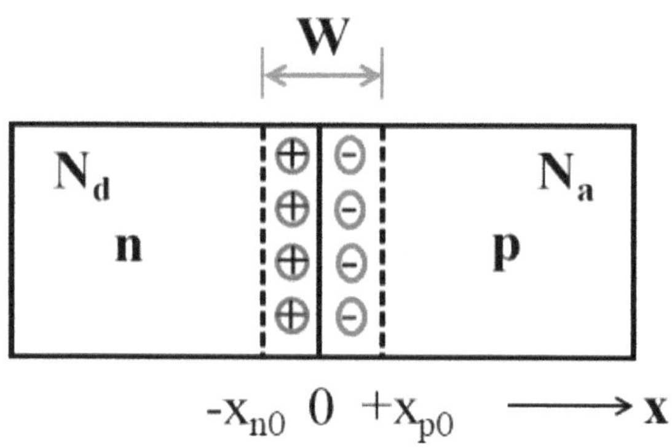


(b) process cross-sectional view

Fig. 7. pn junction diode doped with $N_d = 10^{16}$ donors/㎤ and $N_a = 10^{16}$ acceptors/㎤.

(a) Calculate the contact voltage $V_0$.

$V_0$ is calculated by the following equation.

$$\begin{aligned} V_0 &= \frac{kT}{q}\ln\frac{N_aN_d}{n_i^2} \\ &= 0.0259\times\ln\frac{10^{16}\times10^{16}}{(1.5\times10^{10})^2} \\ &= 0.695\ V \end{aligned} \tag{12}$$

(b) Calculate the depletion width $W$, $x_{p0}$, and $x_{n0}$ when the applied voltage V is 0V across the pn junction diode. At this time, the unit of depletion width is expressed in μm. $\epsilon_{Si}$ = 11.8.

The depletion width W can be calculated as in Eq. (13).

$$\begin{aligned} W &= \sqrt{\frac{2\epsilon_0\epsilon_{Si}(V_0-V)}{q}\left(\frac{1}{N_a}+\frac{1}{N_d}\right)} \\ &= \sqrt{\frac{2\times8.854\times10^{-14}\times11.8\times0.695}{1.6\times10^{-19}}\times\left(\frac{1}{10^{16}}+\frac{1}{10^{16}}\right)} \\ &= 4.26\times10^{-5}cm \\ &= 0.426\mu m \end{aligned} \tag{13}$$

The calculated depletion width $W$ is 0.426 μm.

Also, the depletion widths $x_{n0}$ and $x_{p0}$ can be calculated as in Eqs. (14) and (15).

$$x_{n0} = \frac{N_a}{N_a + N_d} \times W \tag{14}$$
$$= \frac{10^{16}}{10^{16} + 10^{16}} \times 0.426$$
$$= 0.213\ \mu m$$

$$x_{p0} = \frac{N_d}{N_a + N_d} \times W \tag{15}$$
$$= \frac{10^{16}}{10^{16} + 10^{16}} \times 0.426$$
$$= 0.213\ \mu m$$

Since the concentrations of $N_a$ and $N_d$ in the pn junction diode are the same, the depletion widths of $x_{n0}$ and $x_{p0}$ are 0.213 ㎛, respectively.

**(c) Calculate the abrupt junction capacitance $C_j$ in case of junction area A=10 $\mu m^2$.**

Junction capacitance $C_j$ is calculated as

$$C_j = \epsilon_0 \epsilon_{Si} \frac{A}{W} \tag{16}$$
$$= 8.854 \times 10^{-14} \times 11.8 \times \frac{10\mu m^2}{0.426\mu m}$$
$$= 2.45 fF$$

The calculated $C_j$ is 2.45fF.

## 2.2 Forward Bias and Reverse Bias

There are three bias conditions namely equilibrium state, forward bias and reverse bias as shown in Fig. 8 depending on the biasing condition at the pn junction diode. Fig. 8 shows the cross-sectional view of the process, potential diagram, and energy band diagram considering depletion region due to bias voltage.

In the equilibrium state where the terminal of the pn junction is open and the voltage is not applied from the outside as shown in Fig. 8(a), the charge carriers can not move due to the potential barrier formed inside the depletion region of the pn junction diode, and the current does not flow in the pn junction.

Applying the forward bias voltage to the pn junction diode is a case of applying a greater voltage to the p region than to the n region, as shown in Fig. 8(b). When the forward bias voltage $V_F$ is applied to the pn junction in the forward bias, the potential barrier of the pn junction is reduced to $V_0 - V_F$ by $V_F$ as shown in the potential diagram of Fig. 8(b). And electrons which are the minority carriers in the p region, are injected from the n region into the p region, and holes which are the minority carriers in the n region are injected from the p region into the n region. Due to the diffusion of these injected holes and electrons, a current flows through the pn junction diode.

When forward bias voltage is applied to the pn junction diode, many holes in the p region and many electrons in the n regions are injected through the depletion region as shown in Fig. 9. Due to the diffusion of these injected holes and electrons, a current flows through the pn junction diode. Therefore, as holes are injected from p region to n region, some holes recombine with negative ions in the depletion region and some electrons recombine with positive ions in the depletion region as electrons are injected from n region to p region. Thus, the number of negative and positive ions of the pn junction diode is reduced. Therefore, when a forward bias voltage is applied, the reduction of the positive and negative ions in the depletion region causes the depletion width to be reduced.

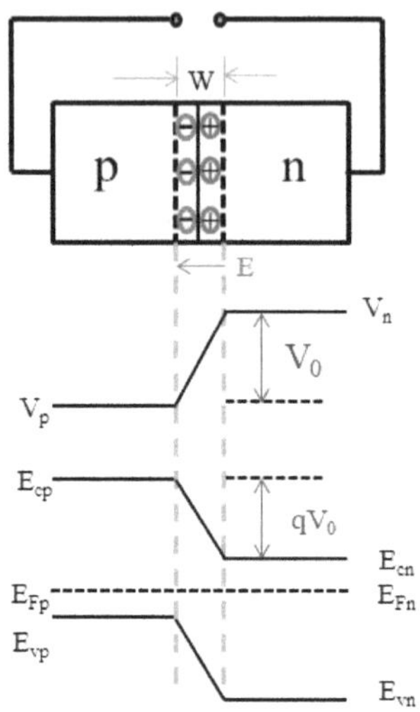


(a) equilibrium state

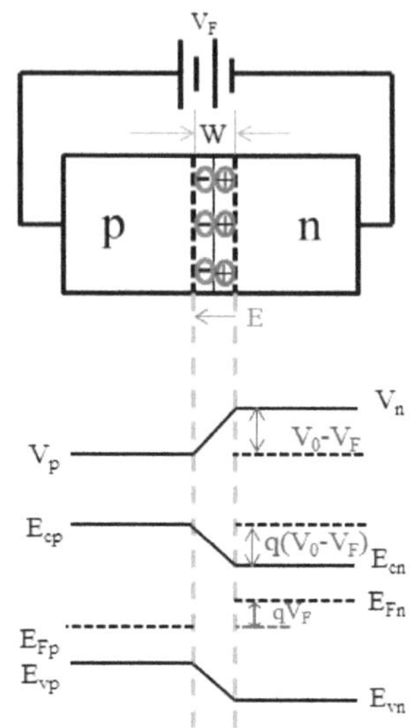


(b) forward bias

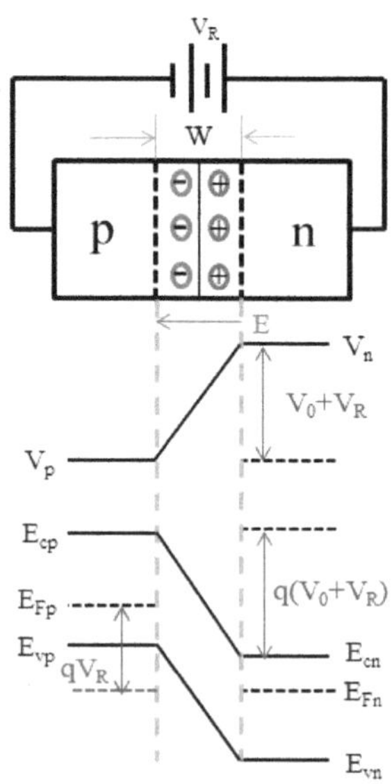


(c) reverse bias

Fig. 8. Process cross-sectional view, potential diagram, and energy band diagram considering depletion region according to bias.

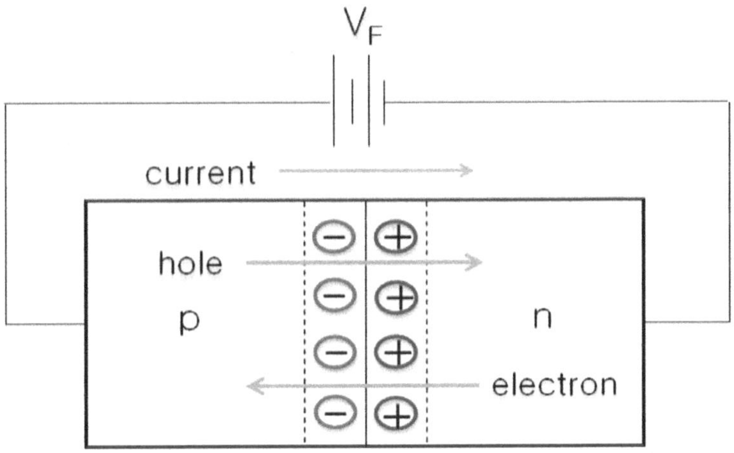


Fig. 9. When the forward bias is applied to the pn junction, the charge carrier movement direction.

The last bias condition of three biasing methods in pn junction diode shown in Fig. 8 is reverse biasing. Applying the pn junction with the reverse bias voltage is a case where the voltage of the n region is larger than the voltage of the p region as shown in Fig. 8(c). When the reverse bias voltage $V_R$ is applied to the pn junction in the reverse bias, as shown in the potential diagram of Fig. 8(c), the potential barrier of the pn junction becomes higher by $V_R$ to $V_0 + V_R$, the injection of holes from the p region into the n region and the injection of electrons into the p region from the n region are not carried out, and currents injected by the majority carriers do not flow. In addition, when a reverse bias voltage is applied to the pn junction, a positive voltage applied to the n region attracts electrons which are majority carriers in the n region, and positive ions are additionally formed in the depletion region of the n region. The negative voltage applied to the p region attracts the holes in the p region, so that negative ions are additionally formed in the depletion region of the p region. In this way, the addition of the positive ions of the n region and the negative ions of the p region mean that the depletion width increases.

If the reverse bias voltage is applied to the pn junction diode, there is a reverse saturation current $I_0$ firstly due to the charge carriers caused by thermally generated EHPs in the depletion region, and secondly due to minority carriers which are electrons that are entered into the depletion region by thermal motion near the depletion region in the p region and holes that are entered into the depletion region by thermal motion near the depletion region in the n region. In the depletion region, the electric field is directed from the positive ions to the negative ions as shown in Fig. 10, so that the holes in the depletion region drift toward the electric field, and the electrons in the depletion region drift in the direction opposite to the electric field. The reverse saturation current of the pn junction diode is dominant due to the thermally generated EHP in the depletion region.

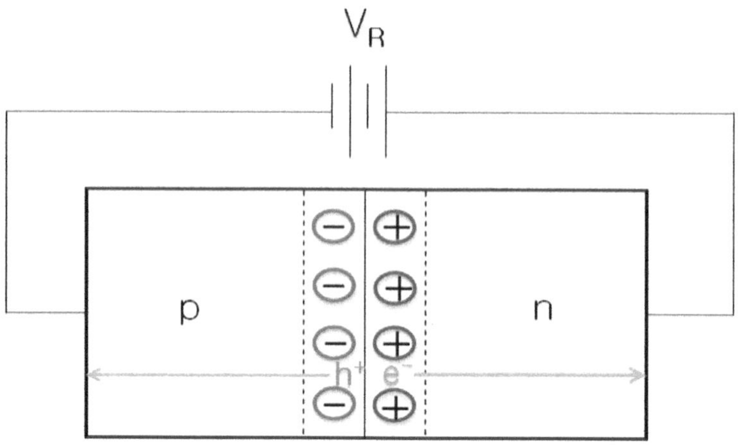


Fig 10. Charge carrier movement direction when reverse bias is applied to pn junction.

**Example 2-1**

*Answer the following questions for a pn junction diode below with the doping concentrations of $N_d = 10^{16}$ donors/㎤ and $N_a = 10^{15}$ acceptors/㎤ at T=300°K.*

(a) Calculate $p_0$ and $n_0$ in the p region at an equilibrium state of V=0 V.

In the case of an equilibrium state of V=0 V, if $p_0$ and $n_0$ are calculated in the p region, $p_0$ is almost equal to $N_a$, so $p_0$ is as follows.

$$p_0 \cong N_a = 10^{15}\ holes/cm^3 \tag{17}$$

$n_0$ is calculated from the relation $n_0 p_0 = n_i^2$, and $n_0$ is as follows.

$$n_0 = \frac{n_i^2}{p_0} = \frac{(1.5 \times 10^{10})^2}{10^{15}} = 2.25 \times 10^5\ electrons/cm^3 \tag{18}$$

(b) Calculate $n_0$ and $p_0$ in the n region at an equilibrium state.

$n_0$ in the n region is almost equal to $N_d$, so $n_0$ is as follows.

$$n_0 \cong N_d = 10^{16}\ electrons/cm^3 \tag{19}$$

And $p_0$ is calculated from the relationship $n_0 p_0 = n_i^2$, $p_0$ can be obtained from the following equation.

$$p_0 = \frac{n_i^2}{n_0} = \frac{(1.5 \times 10^{10})^2}{10^{16}} = 2.25 \times 10^4 \ holes/cm^3 \quad (20)$$

(c) Calculate the Fermi levels in the p and n regions: $E_{ip} - E_{Fp}$ and $E_{Fn} - E_{in}$.

In the p region, $E_{ip} - E_{Fp}$ can be used by the following equation.

$$p_0 = n_i \times e^{(E_{ip} - E_{Fp})/kT} \quad (21)$$

Passing $n_i$ to the left side of the equation gives the following equation.

$$\frac{p_0}{n_i} = e^{(E_{ip} - E_{Fp})/kT} \quad (22)$$

Taking the logarithm of both sides of the equation gives the following equation:

$$\ln \frac{p_0}{n_i} = \frac{E_{ip} - E_{Fp}}{kT} \quad (23)$$

Expand this expression for $E_{ip} - E_{Fp}$ and substitute the corresponding parameter values as follows:

$$E_{ip} - E_{Fp} = kT \times \ln \frac{p_0}{n_i} = 0.0259 \times \ln \frac{10^{15}}{1.5 \times 10^{10}} = 0.288 \ eV \quad (24)$$

So, $E_{ip} - E_{Fp}$ is 0.288 eV.

Next, calculate $E_{Fn} - E_{in}$ in the n region. In the n region, $E_{Fn} - E_{in}$ can be used by the following equation.

$$n_0 = n_i \times e^{(E_{Fn} - E_{in})/kT} \tag{25}$$

Passing $n_i$ to the left side of the equation gives the following equation.

$$\frac{n_0}{n_i} = e^{(E_{Fn} - E_{in})/kT} \tag{26}$$

Taking the logarithm of both sides of the equation gives the following equation:

$$\ln\frac{n_0}{n_i} = \frac{E_{Fn} - E_{in}}{kT} \tag{27}$$

Expand this expression for $E_{Fn} - E_{in}$ and substitute the corresponding parameter values as follows:

$$E_{Fn} - E_{in} = kT \times \ln\frac{n_0}{n_i} = 0.0259 \times \ln\frac{10^{16}}{1.5 \times 10^{10}} = 0.347\ eV \tag{28}$$

So, $E_{Fn} - E_{in}$ is 0.347 eV.

(d) Calculate the contact voltage $V_0$.

$V_0$ can be obtained by expanding as shown below.

$$V_0 = \frac{kT}{q} ln \frac{N_a \cdot N_d}{n_i^2} \tag{29}$$

$$= 0.0259 \times \ln \frac{10^{15} \cdot 10^{16}}{(1.5 \times 10^{10})^2}$$

$$= 0.635\ V$$

**(e) Draw a potential diagram and an energy band diagram at an equilibrium state.**

As can be seen from Fig. 11, we can see that the voltage $V_n$ in the n region is 0.635 V higher than the voltage $V_p$ in the p region. On the other hand, the energy band diagram shows that the conduction energy level $E_{cp}$ of p region is 0.635 eV by q $V_0$ greater than $E_{cn}$ of n region.

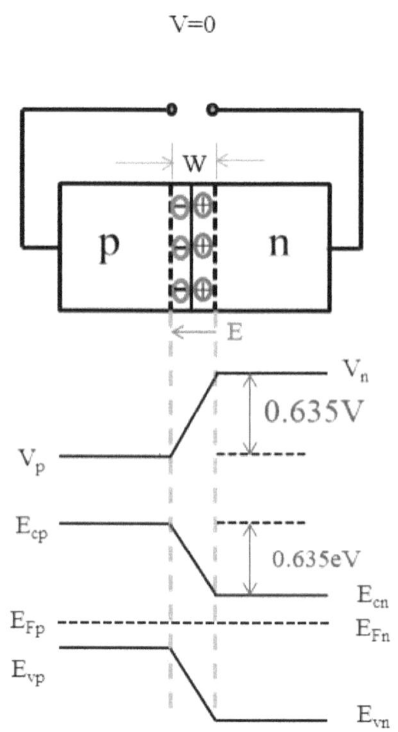


Fig. 11. Potential diagram and energy band diagram at an equilibrium state.

(f) Draw a potential diagram and an energy band diagram at a forward bias state of V=0.3 V.

The next problem is to draw the potential diagram and the energy band diagram under forward bias as shown in Fig. 12. As can be seen in the potential diagram with forward bias of $V_F$ = 0.3 V, the voltage $V_n$ in the n region is 0.335 V higher than the voltage $V_p$ in the p region, which is $V_0 - V_F$. In case of forward biasing, potential barrier is lowered by $V_F$ than equilibrium. On the other hand, it can be seen that the conduction energy level $E_{cp}$ in the p region is greater by 0.335 eV, q( $V_0 - V_F$), which is reduced by q $V_F$, than the $E_{cn}$ in the n region.

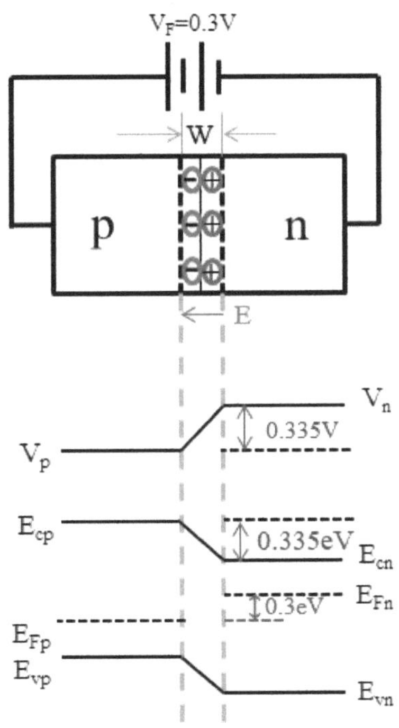


Fig. 12. Potential diagram and energy band diagram at a forward bias.

(g) Draw a potential diagram and an energy band diagram at a reverse bias state V=−0.2 V.

The next problem is to draw the potential diagram and the energy band diagram in the reverse bias state as shown in Fig. 13. As can be seen from the potential diagram at the reverse bias state of $V_R$ = 0.2 V, the voltage $V_n$ in the n region is 0.835 V higher than the voltage $V_p$ in the p region, which is $V_0 + V_R$. In the case of biasing in the reverse direction, it can be seen that the potential barrier is higher by $V_R$ than equilibrium. On the other hand, the energy band diagram shows that the energy level of the electrons in the n region is lower than the p region by q$V_R$, and $E_{Fn}$ is $qV_R$ below the $E_{Fp}$. It can be seen that the conduction energy level $E_{cp}$ of p region is greater by 0.835 eV, which is q($V_0 + V_R$), which is increased by q$V_R$, than $E_{cn}$ of n region.

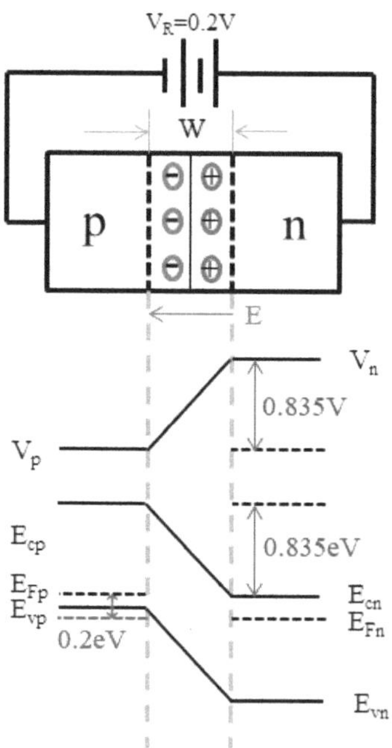


Fig. 13. Potential diagram and energy band diagram at a reverse bias.

## 2.3 I-V Characteristic Curve

The cross-sectional view and symbol of the pn junction diode are shown in Fig. 14. The p and n region are called anode and cathode, respectively.

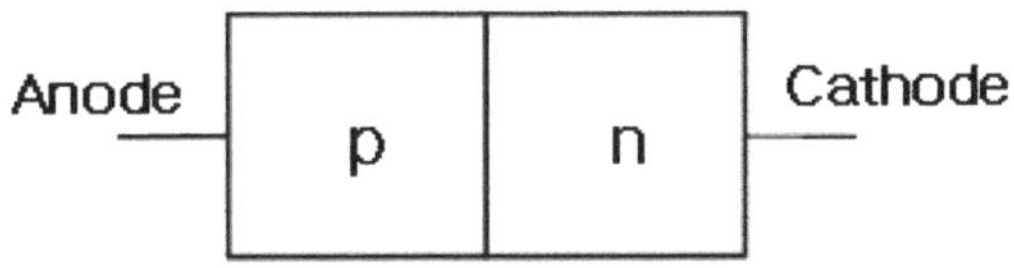


(a) process cross-sectional view

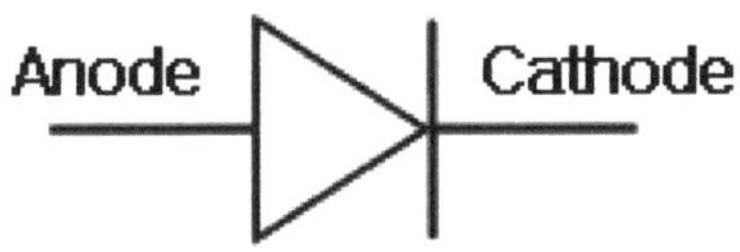


(b) symbol

Fig. 14. pn junction diode.

Fig. 15 shows the current I flowing from the anode to the cathode of the pn junction diode when voltage V is applied across the diode. As shown in Fig. 15, when the voltage V is applied across the diode, the curve of the diode current I is shown in Fig. 16. In the reverse-biased case, the current scale was enlarged. In the forward bias where the applied voltage V is positive in the I-V curve of the diode, the diode current I increases exponentially with respect to the applied voltage V. On the other hand, in the $V_{br}$ voltage section where the applied voltage V is negative between 0 V and breakdown voltage $V_{br}$, the reverse saturation current $I_0$ flows, and when the reverse voltage is applied above $V_{br}$, infinite current flows from the n region to the p region.

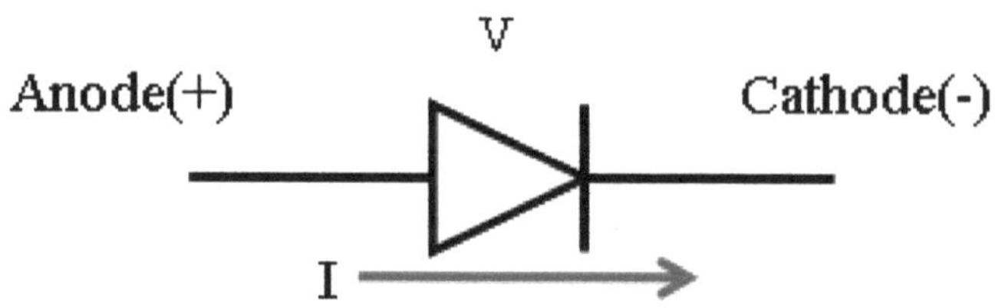


Fig. 15. Definition of the current I flowing when the voltage V is applied to the diode.

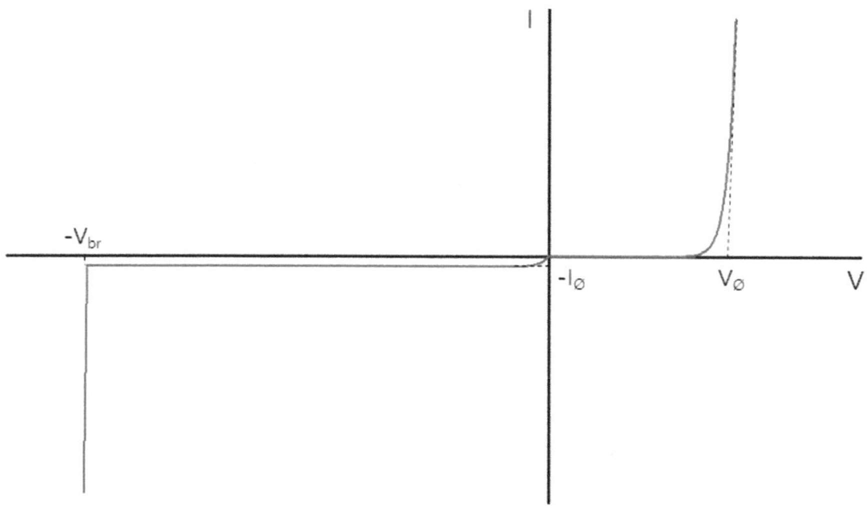


Fig. 16. I-V characteristic curve of pn junction diode.

When the voltage V is applied to the pn junction diode, the current I flowing through the diode is expressed by Eq. (30).

$$I = I_0 (e^{\frac{V}{V_T}} - 1) \tag{30}$$

where $I_0$ is the reverse saturation current of the pn junction diode at absolute temperature T and $V_T$ is the thermal voltage.

Since the exponential term of Eq. (30) is much larger than 1 when the applied voltage V is a positive voltage, and V is several times higher than $V_T$, the forward current I can be written as in Eq. (31). Therefore, the forward current increases exponentially with the forward bias voltage V. The reverse current I shows the approximated current when the forward voltage V is applied at 3 $V_T$.

$$I \cong I_0\, e^{\frac{V}{V_T}} \tag{31}$$

The exponential term of Eq. (30) is negligibly smaller than 1 when the applied voltage V is a negative voltage and |V| is several times higher than $V_T$, and the reverse current I is approximated as Eq. (32).

Therefore, the reverse current flows from the n region to the p region by the reverse saturation current $I_0$. It should be noted that this reverse saturation current is very small but proportional to the junction area of the pn junction diode and increases with increasing temperature. The reverse current I shows the approximated current when the voltage V is applied at -3 $V_T$.

$$I \cong -I_0 \tag{32}$$

Since $I_0$ and $V_T$ are functions of temperature, diode current I changes with T as shown in Fig. 17 when forward voltage V is applied. When the diode current is constant, the voltage drop of the diode decreases by about 2 mV as the temperature increases by 1℃. The voltage change of the diode with temperature is applied to the reference voltage generator circuit and the thermometer circuit design.

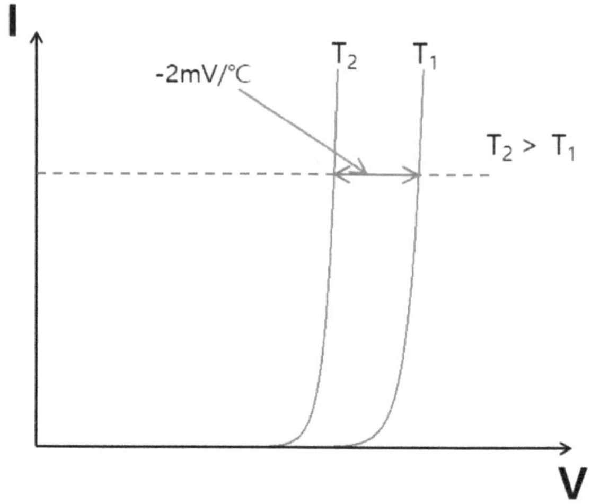


Fig. 17. Effect of voltage drop due to temperature increase when a diode is applied with forward bias.

## 2.4 Breakdown Mechanism

As shown in Fig. 16, when reverse bias voltage is applied above breakdown voltage $V_{br}$, junction breakdown occurs and large current flows suddenly into the p region from the n region. Reverse breakdown mechanisms in the pn junction diodes include Zener breakdown and avalanche breakdown.

Zener breakdown occurs as shown in Fig. 18(b) even when the energy band has a low reverse bias voltage of several volts in case of the heavily doped pn junction diode. If the depletion width of the pn junction diode is very narrow, the electrons in the valence band of the p region form a reverse current due to the Zener effect of tunneling to the conduction band of the n region.

The direction of electron tunneling occurs from the valence band in the p region to the conduction band in the n region as shown in Fig. 18(b). The tunneling distance W in Fig. 18(b) becomes narrower as impurities are added at higher concentration and reverse bias is larger. The tunneling distance W is the gap between the valence band edge of the p region and the conduction band edge of the n region as shown in Fig. 18(b). When a reverse voltage above $V_{br}$ is applied, an infinite current flows from the n region to the p region due to the electron tunneling.

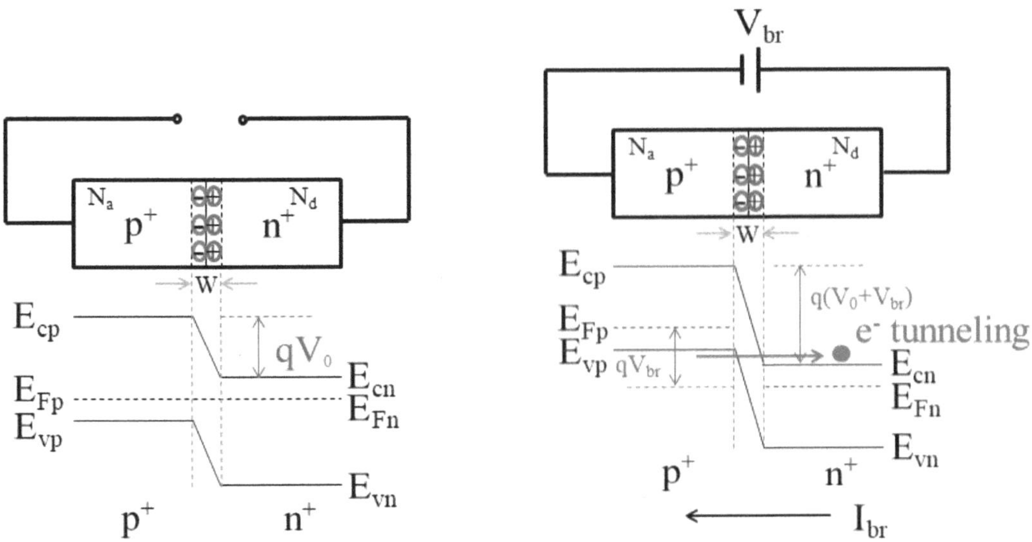


(a) equilibrium state (b) when reverse bias is applied

Fig. 18. Zener effect.

## Example 2-3

*Draw an energy band diagram and a picture depicting the electron tunneling when applying a reverse voltage of $V_{br}$ or higher to a pn junction diode injected with high concentrations of impurities.*

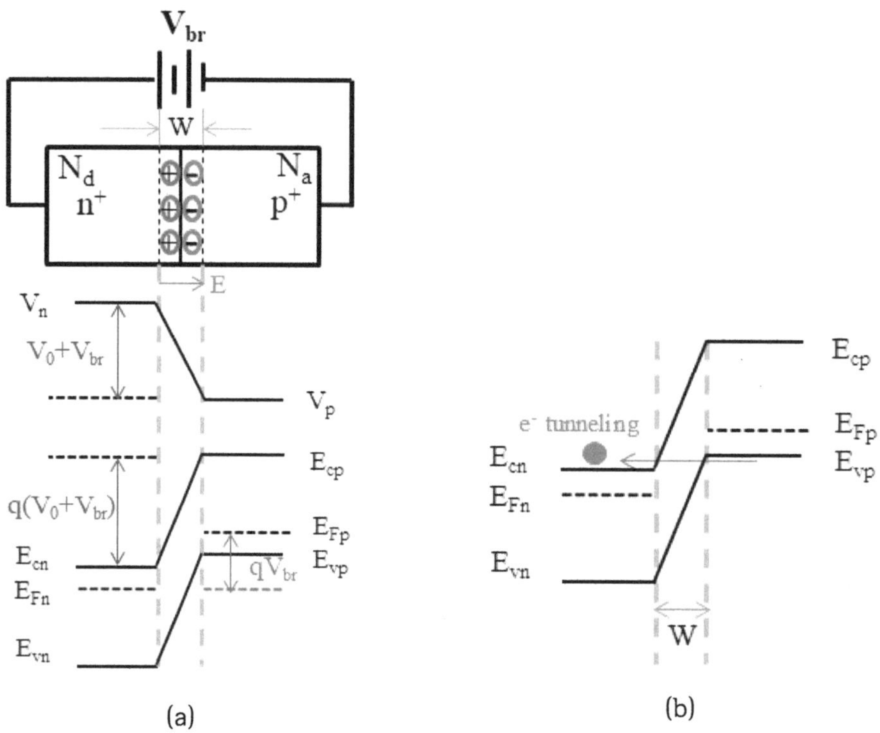


Fig. 19. Picture showing energy band diagram and electron tunneling when a voltage higher than $V_{br}$ is applied to a pn junction diode injected with high concentration of impurities.

The n region of the pn junction diode is located on the left side and the p region is located on the right side. When applying the reverse voltage above $V_{br}$, if you draw the energy band diagram, as shown in Fig. 19(a), $E_{Fp}$ goes up q$V_{br}$ above $E_{Fn}$ and $E_{vp}$ is q($V_0 + V_{br}$) above $E_{vn}$. As shown in Fig. 19(a), the depletion width W is narrow and the edge of the valence band of the p region and the edge of the conduction band of the n region face each other. When a pn junction diode injected with high concentrations of impurities has a reverse bias of $V_{br}$ or more, electrons in the valence band of p region tunnel into the conduction band of n region as shown in Fig. 19(b).

Tunneling of electrons at lightly doped pn junction diode is negligible because of the large depletion width, and avalanche breakdown occurs by impact ionization. If the electric field intensity E in the depletion region is large, the electrons entering from the p region can accelerate to a sufficiently large kinetic energy and cause an ionization collision with the lattice. The EHPs produced by the impact ionization interact with another lattice to cause an avalanche process that proliferates the carriers. The electrons generated by the original electrons and impact ionization are swept into the n region, and the generated holes are swept into the p region. When a reverse voltage above $V_{br}$ is applied, an infinite current flows from the n region to the p region due to the avalanche breakdown.

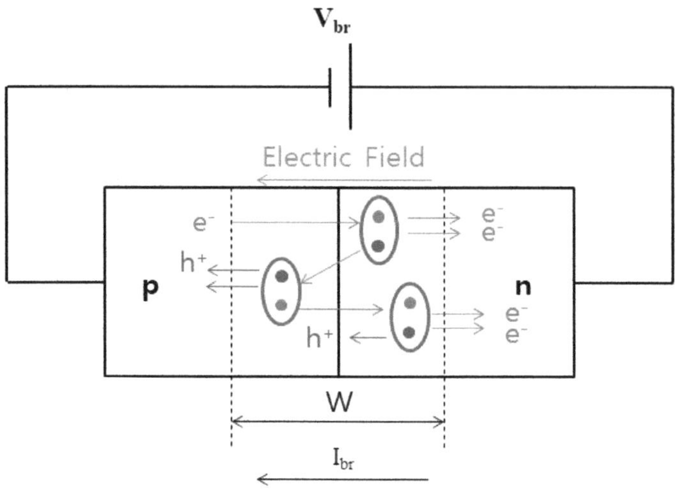


Fig. 20. Avalanche breakdown.

## Exercises

1. Answer the following questions about silicon pn junction diodes with doping concentrations of $N_a = 10^{16}$ acceptors/㎤ and $N_d = 10^{16}$ acceptors/㎤ at T=300°K.

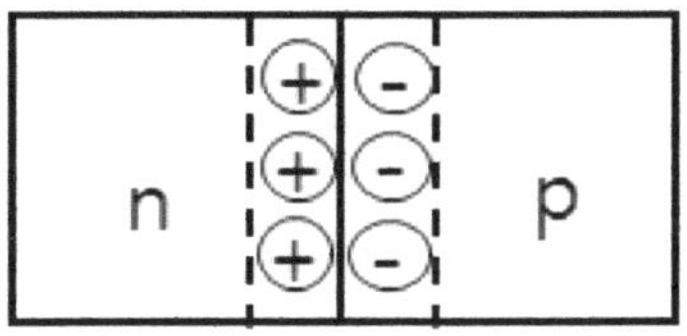


(a) Calculate contact voltage $V_o$.

(b) Draw the potential diagram with the forward bias voltage $V_F$ applied.

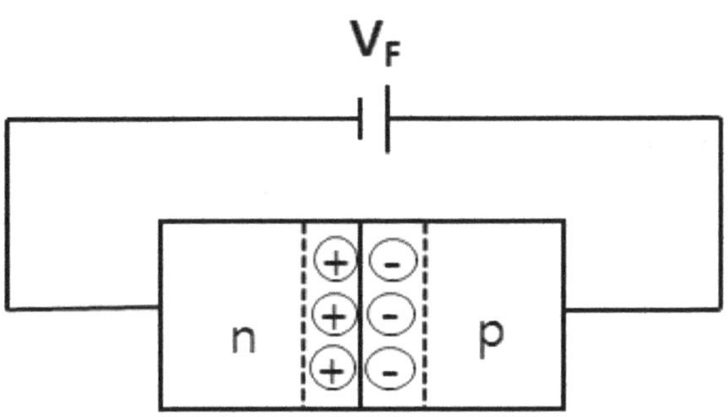

(c) Draw the energy band diagram with the reverse bias voltage $V_R$ applied.

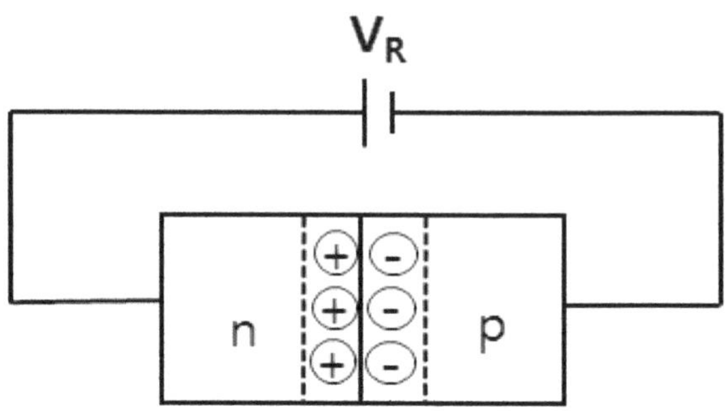


2. Answer the following questions.

(a) Write diode equation for pn junction diode.

(b) Draw the I-V curve of the pn junction diode and mark $V_o$, $I_o$, and $V_{br}$.

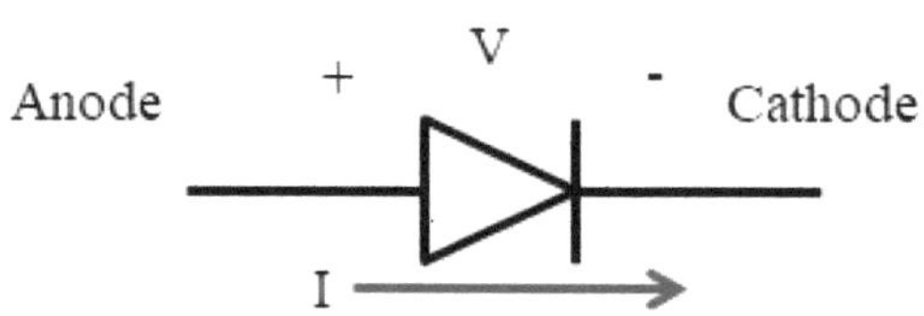


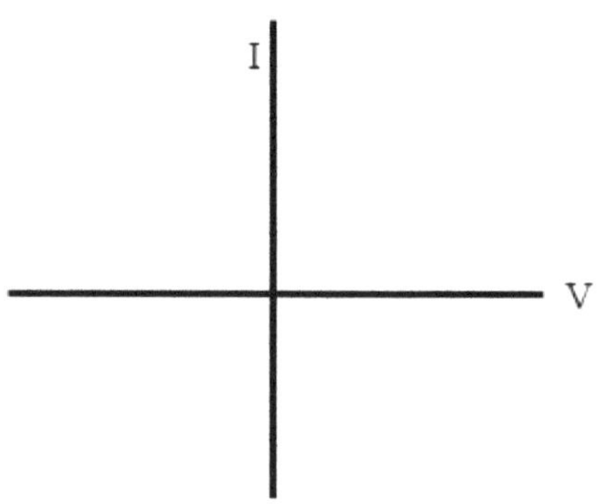

(c) Draw an energy band diagram in the reverse bias state where pn junction diodes doped with impurities at high concentration have electron tunneling from p-type to n-type by Zener breakdown.

(d) Draw an energy band diagram and a picture where an electron tunneling occurs when a reverse voltage of $V_{br}$ or higher is applied to a pn junction diode that is doped with a high concentration.

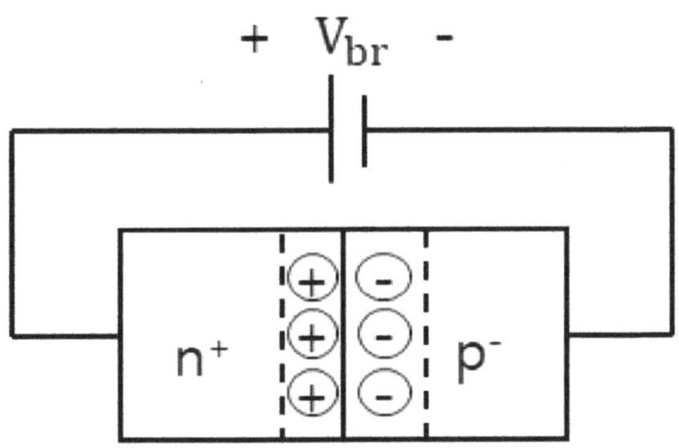


(e) Is Si an extrinsic semiconductor or an intrinsic semiconductor?

3. Answer the following questions about Si pn junction diodes with doping concentrations of $N_a = 10^{15}$ acceptors/㎤ and $N_d = 10^{15}$ donors/㎤ at T=300°K.

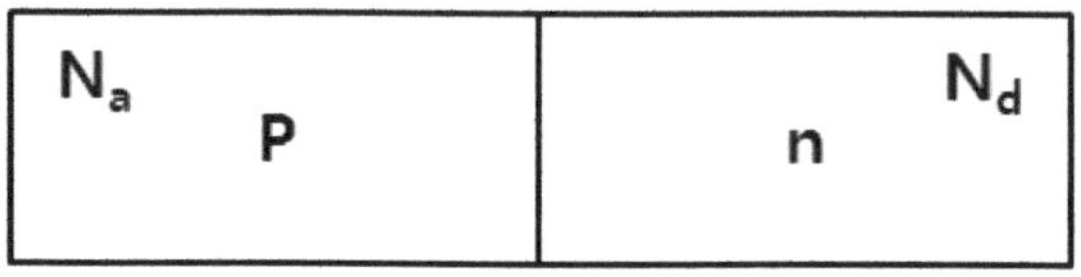


(a) Calculate contact voltage $V_O$.

(b) Draw the potential diagram under the following bias condition.

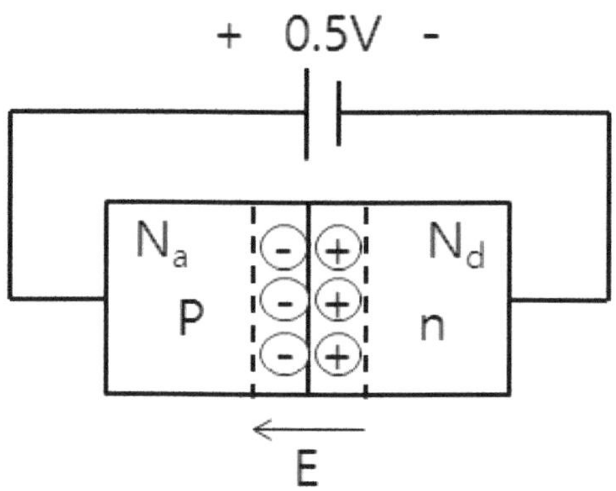


4. Indicate whether the next bias states are forward bias or reverse bias.

(a)

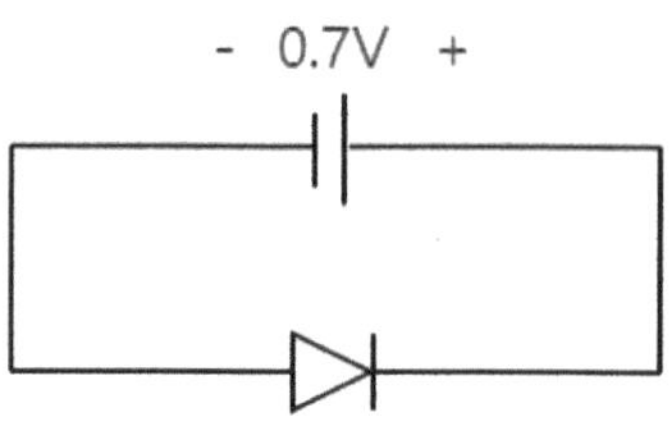


(b)

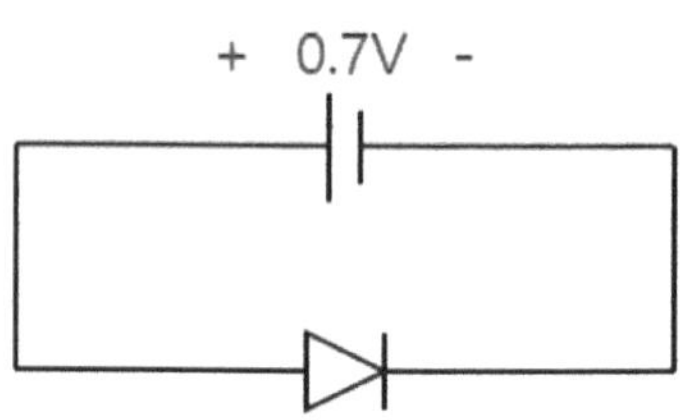

5. Answer the following questions for the Si pn junction diode shown below, with a doping concentration of $N_d = 10^{17}$ donors/㎤ and $N_a = 10^{15}$ acceptors/㎤ at T=300°K. Here, A = $1\,\mu\mathrm{m}^2$.

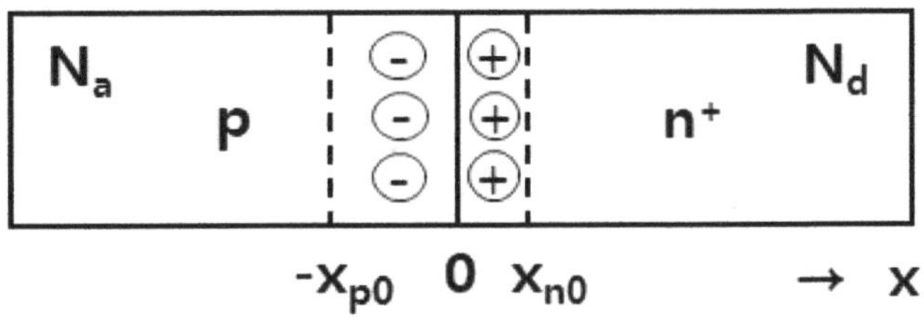


(a) Calculate the contact voltage $V_0$.

(b) Draw the potential diagram in the equilibrium state.

(c) Draw the energy band diagram in the equilibrium state.

(d) Draw the charge density curve according to x.

(e) Calculate the depletion width W, $x_{n0}$, and $x_{p0}$. The unit should be $\mu$m. Here, $\epsilon_{Si}$ = 11.8.

$$W = [\frac{2\epsilon_0 \epsilon_{Si} V_0}{q}(\frac{1}{N_a} + \frac{1}{N_d})]^{\frac{1}{2}}$$

6. Answer the following questions for the Si pn junction diode shown below, with a doping concentration of $N_d = 10^{15}$ donors/㎤ and $N_a = 10^{17}$ acceptors/㎤ at T=300°K.

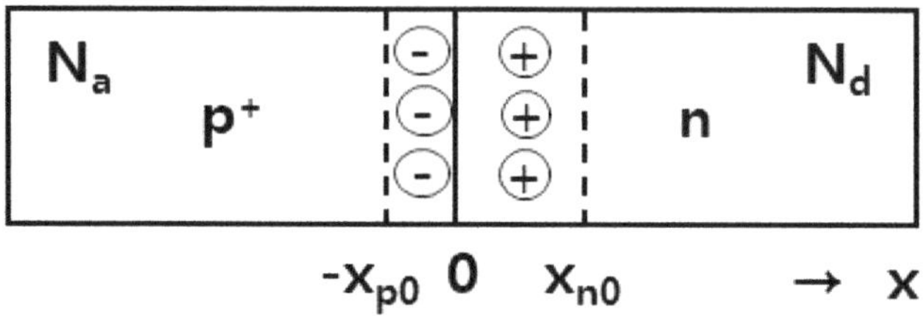


(a) Calculate the positions of the Fermi levels $E_{ip}$-$E_{Fp}$ and $E_{Fn}$-$E_{in}$ in the p and n regions.

(b) Calculate the contact voltage $V_0$.

(c) Draw the potential diagram in the equilibrium state.

(d) Draw the energy band diagram in the equilibrium state.

(e) Draw the charge density curve according to x.

(f) Calculate the depletion width W, $x_{n0}$, and $x_{p0}$. The unit should be $\mu$m.

7. Answer the following questions for the Si pn junction diode shown below, with a doping concentration of $N_d = 10^{17}$ donors/㎤ and $N_a = 10^{17}$ acceptors/㎤ at T=300°K.

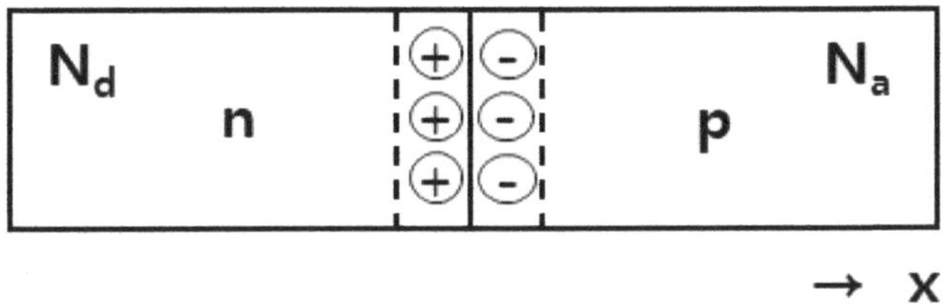


(a) Calculate $p_0$ and $n_0$ in the p region in the equilibrium state.

(b) Calculate $p_0$ and $n_0$ in the n region in the equilibrium state.

(c) Calculate the positions of the Fermi levels $E_{ip}-E_{Fp}$ and $E_{Fn}-E_{in}$ in the p and n regions.

(d) Calculate the contact voltage $V_0$.

(e) Draw the potential diagram in the equilibrium state..

(f) Draw the energy band diagram in the equilibrium state.

8. Answer the following questions for the Si pn junction diode shown below, with a doping concentration of $N_d = 10^{17}$ donors/㎤ and $N_a = 10^{17}$ acceptors/㎤ at T=300°K.

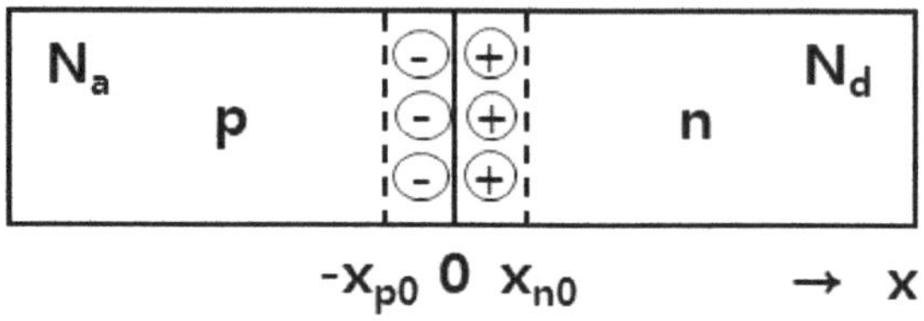


(a) Calculate the contact voltage $V_0$.

(b) Calculate the depletion width W, $x_{n0}$, and $x_{p0}$. The unit should be μm. Here, $\epsilon_{Si}$ = 11.8.

9. Answer the following questions for the Si pn junction diode shown below, with a doping concentration of $N_d = 10^{16}$ donors/㎤ and $N_a = 10^{16}$ acceptors/㎤ at T=300°K.

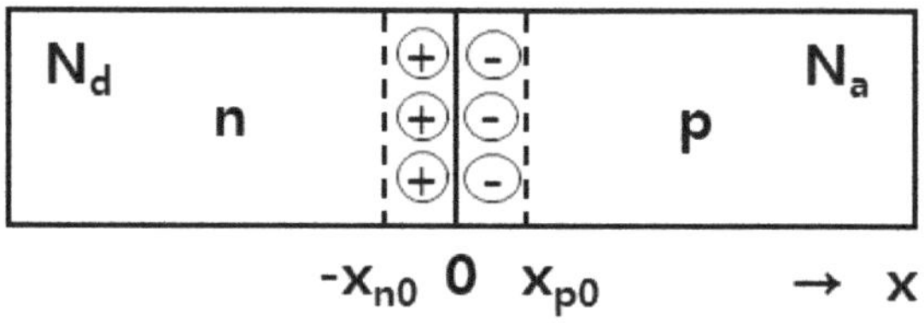


(a) Calculate the contact voltage $V_0$.

(b) Calculate the depletion width W, $x_{n0}$, and $x_{p0}$. The unit should be μm. Here, $\epsilon_{Si}$ = 11.8.

(c) If junction area A = $10\mu m^2$, calculate junction capacitance $C_j$.

10. Answer the following questions about pn junction diode.

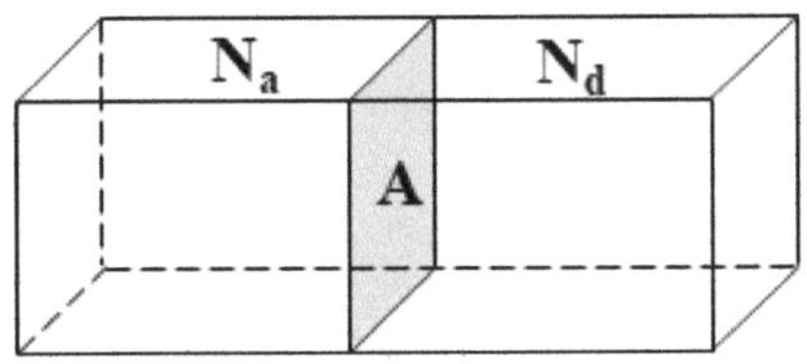


Here, $N_a = 10^{16}$ acceptors/㎤, $N_d = 10^{16}$ donors/㎤, A = 10 $\mu\text{m}^2$ , $\epsilon_{Si}$ = 11.8, and T=300°K.

(a) Calculate the contact voltage $V_0$.

(b) Calculate the depletion width W.

(c) Calculate the junction capacitance $C_j$.

## Chap. 2 References

[1] Sanghee Son et al., Electronic Circuits, First Book, 2015.

[2] Cheolwoo Kim et al., Microelectronics, Hanti Media, 2012.

[3] Kyungwook Shin, Electronic Circuits: Core Concepts to Applications, Hanbit Academy, 2014.

[4] Weonseop Jeong et al., Microelectronics, Hanti Media, 2011.

[5] Gyedal Kwak et al., Solid-State Electronics, Pearson, 2009.

[6] B. G. Streetman, Solid State Electronic Devices, Prentice-Hall International, 1990.

# 3 CHAPTER

# MOS Devices

## 3.1 Device Structure of MOS Transistors

MOSFET (Metal-Oxide-Semiconductor Field-Effect Transistor) is a semiconductor device most commonly used in IC (Integrated Circuit) design, which is a circuit fabricated on a silicon chip. The structure of the MOSFET is made by using Si as the semiconductor, $SiO_2$ as the oxide between the semiconductor and the metal, and metal as the gate electrode. The name of the MOSFET comes from the physical structure of the device, and the poly-silicon gate deposited with metallic material is recently used instead of metal.

In general, MOS transistors can be made very small compared to BJTs, and they are simple to manufacture and operate at low power, so they are used in semiconductor chip designs such as CPUs, memories, digital ICs, and analog ICs.

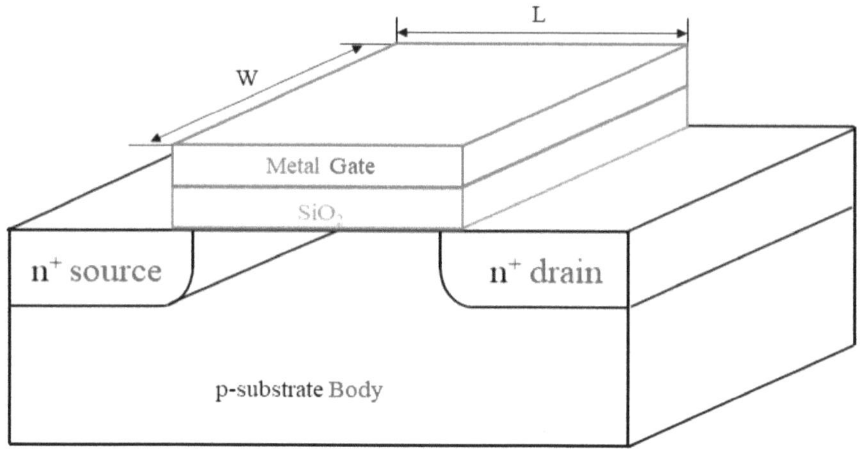


(a) physical structure

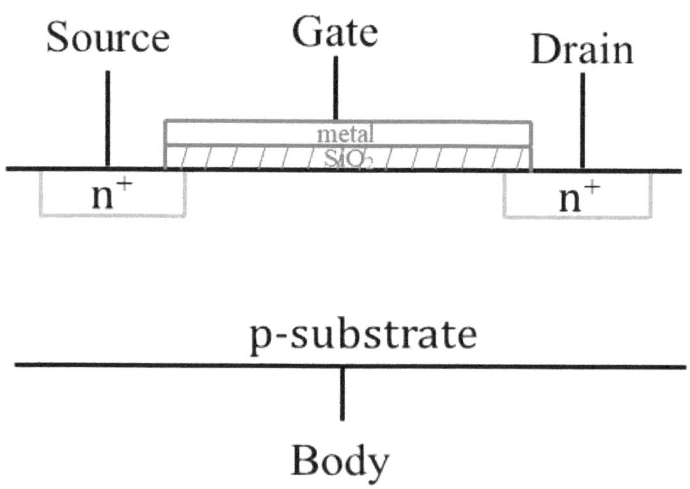


(b) process cross-sectional view

Fig. 1. Enhancement mode NMOS transistor.

Fig. 1 shows the physical structure and process cross-sectional view of an enhancement mode NMOS transistor. An NMOS transistor is formed on a p-substrate. On the surface of the p-substrate, a $SiO_2$ layer, which is a gate oxide with good electrical characteristics, is formed, and a metal gate is formed on the gate oxide. In Fig. 1, the n+ source and drain region are made by injecting donor impurities at a high concentration into a relatively low-doped p-substrate. In Fig. 1(b), one of the n+ diffused regions is a source and the other is a drain. Meanwhile, in Fig. 1(b), p-substrate is the body. The metal gate overlaps the n+ source and drain regions. An NMOS transistor is a 4-terminal device of source, drain, gate, and body, and no current flows from the drain to the source unless an n-channel is formed between the source and the drain. In Fig. 1, W is the channel width and L is the channel length.

The top of Fig. 2(a) shows the cross-sectional view of NMOS transistor and the bottom of Fig. 2(a) shows the energy band diagram in the channel direction of the NMOS transistor in the equilibrium state. As shown in the top of Fig. 2(a), each depletion region at the source junction and drain junction are mainly formed in the substrate which is the lightly doped region.

In the process cross-sectional view, parasitic pn junction diodes are formed in the source junction and drain junction, and in order for the MOS device to operate normally, a bias must be applied in the reverse direction to turn off these pn junction diodes. Therefore, in many analog and digital IC designs, the body voltage of the NMOS transistor is applied with a voltage equal to or lower than the minimum voltage applied to the drain and source nodes. In general, it can be considered that GND is applied to the body voltage of the NMOS transistor.

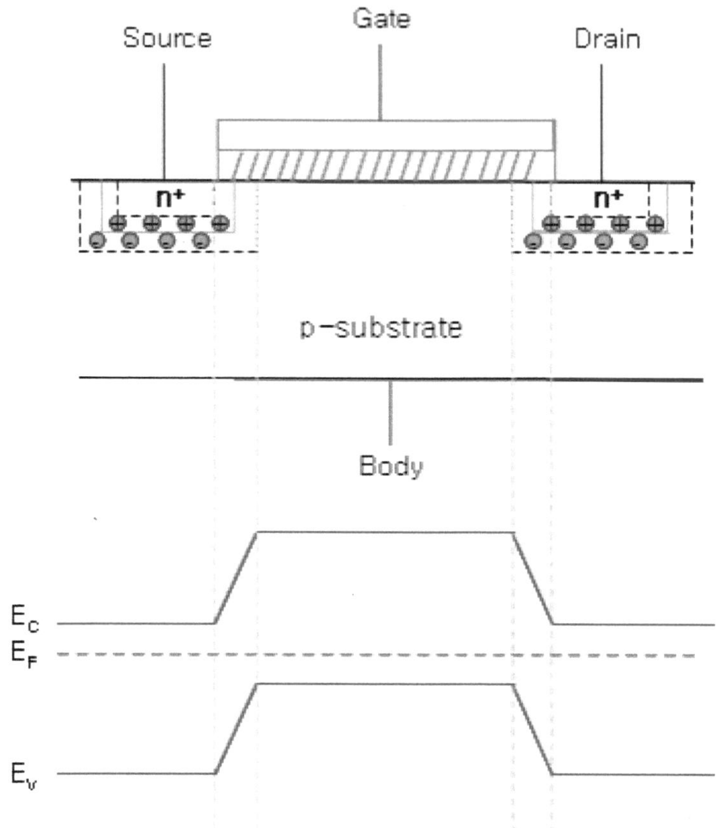


(a)

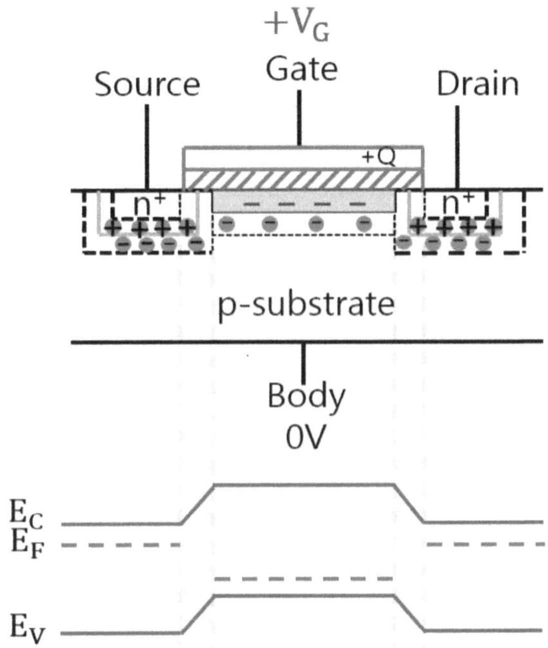


(b)

Fig. 2. The energy band diagram in the channel direction of the enhancement-type NMOS transistor in the equilibrium state.

When the gate voltage of the NMOS transistor is 0 V, the energy band diagram according to the channel is flat at the equilibrium state and the Fermi level at the n+ source and drain regions is near the conduction energy level as shown in the bottom figure of Fig. 2(a).

Since the Fermi level in the substrate is located near the valence energy level, there is a potential barrier for electrons passing from the source to the drain, so no current flows. Due to the potential barrier where the potential of the n+ source region is higher than that of the p-substrate, any electrons in the n+ source region cannot be injected from the source to the p-substrate.

If a positive voltage is applied to the gate while applying a voltage of 0V to the p-substrate, which is the body of the NMOS transistor, positive charges are attached to the metal gate. In response to this, a depletion region in which acceptor impurities are ionized is formed first on the p-substrate under the gate oxide, and an inversion layer is formed on the silicon surface as shown in Fig. 2(b). These electrons form the channel of the MOSFET and allow current to flow from the drain to the source.

As shown in Fig. 2, the effect of the gate voltage is to change the conductivity of the channel for a low drain-source voltage, $V_{DS}$. If the gate voltage is applied above the threshold voltage $V_T$, the conduction energy level of the p-substrate under the gate oxide moves closer to the Fermi level. Thus the potential barrier formed at the source junction decreases. If the potential barrier is sufficiently reduced because the gate voltage is applied above the threshold voltage $V_T$, electrons will move from the source to the drain. Therefore, the gate voltage of the MOS transistor controls the $I_{DS}$, which is the current flowing from the drain to the source.

$V_T$ is the minimum gate voltage required to induce the channel. As shown in Fig. 3, the $V_{GS}$ voltage of the NMOS transistor is greater than $V_T$ so that the n-channel is induced under the gate oxide. Here, the $V_{GS}$ voltage is the voltage difference between the gate voltage and the source voltage. In this way, when the $V_{GS}$ voltage is 0 V, it is in a normally off state. Since a sufficient positive gate voltage is applied, the transistor which induces the n-channel is called an enhancement NMOS transistor. On the other hand, when the $V_{GS}$ voltage is 0V, a transistor that already has an n-channel and requires a negative gate voltage to turn off the device is called a depletion mode NMOS transistor. A depletion mode MOS transistor is a device that is normally on. In general, enhancement mode MOS transistors are often used, so enhancement mode is omitted and called MOS transistors. Depletion mode MOS transistors, on the other hand, are used in some cases for analog circuit designs, and they must be in depletion mode. Fig. 3 shows the symbol and process cross-sectional view of the NMOS transistor.

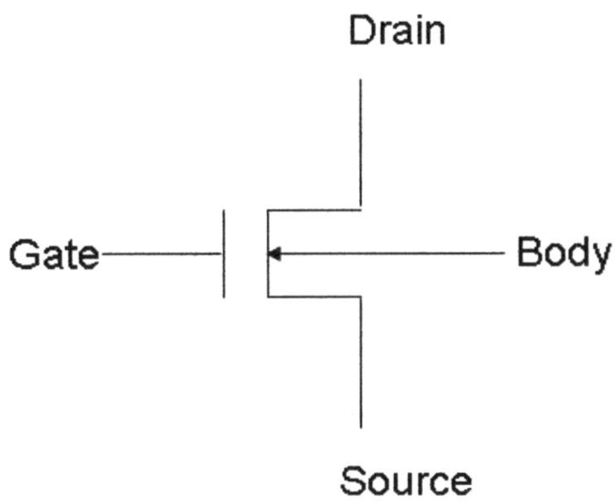


(a) symbol

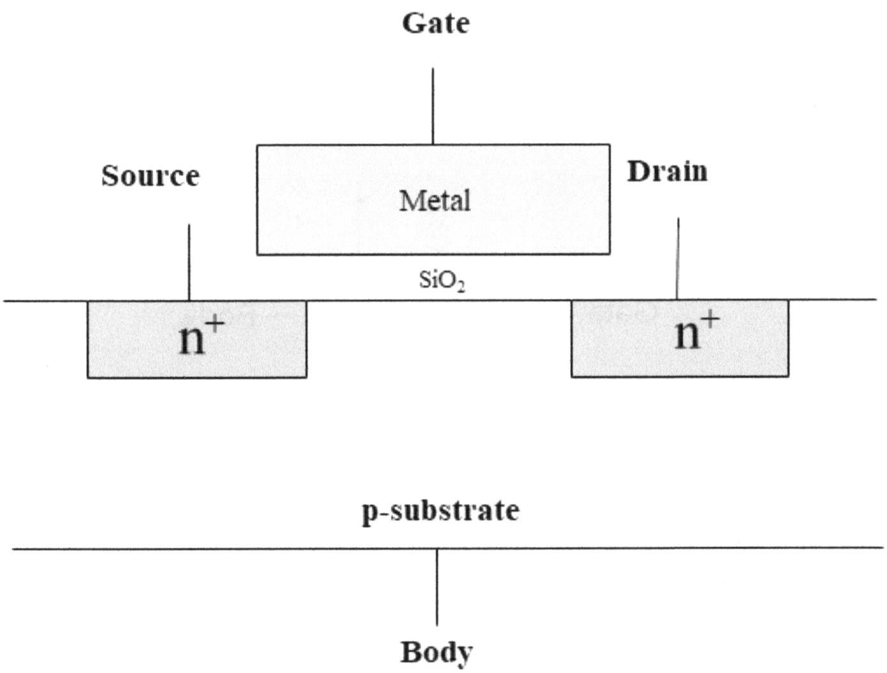


(b) process cross-sectional view

Fig. 3. NMOS transistor.

Fig. 4 shows the symbol and process cross-sectional view of the PMOS transistor. A PMOS transistor is formed on a N-Well in the p-substrate. First, N-Well is formed in the p-substrate. After the N-Well is formed, n+ diffusion for N-Well pick-up is formed. Next, on the surface of the N-Well, a $SiO_2$ layer, which is a gate oxide with good electrical characteristics, is formed, and then a metal gate is formed on the gate oxide. After the metal gate is formed, the p+ source and drain region are made by injecting acceptor impurities at a high concentration into a relatively low-doped N-Well. In Fig. 4(b), one of the p+ diffused regions is a source and the other is a drain. Meanwhile, in Fig. 4(b), the n+ diffused region formed in the N-Well is the body. The metal gate overlaps the p+ source and drain regions. A PMOS transistor is a 4-terminal device of source, drain, gate, and body, and no current flows from the source to the drain unless an p-channel is formed between the source and the drain.

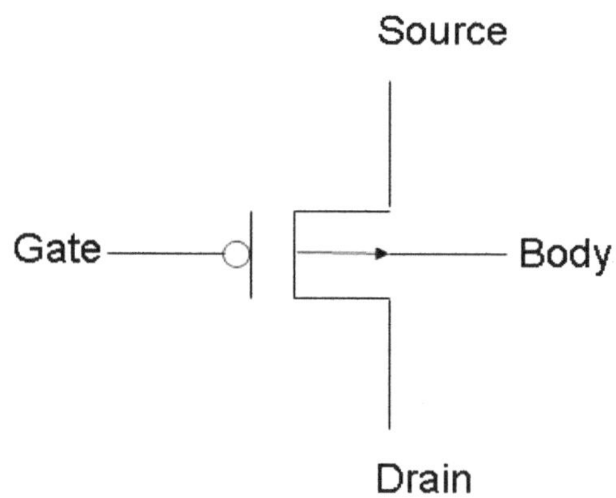


(a) symbol

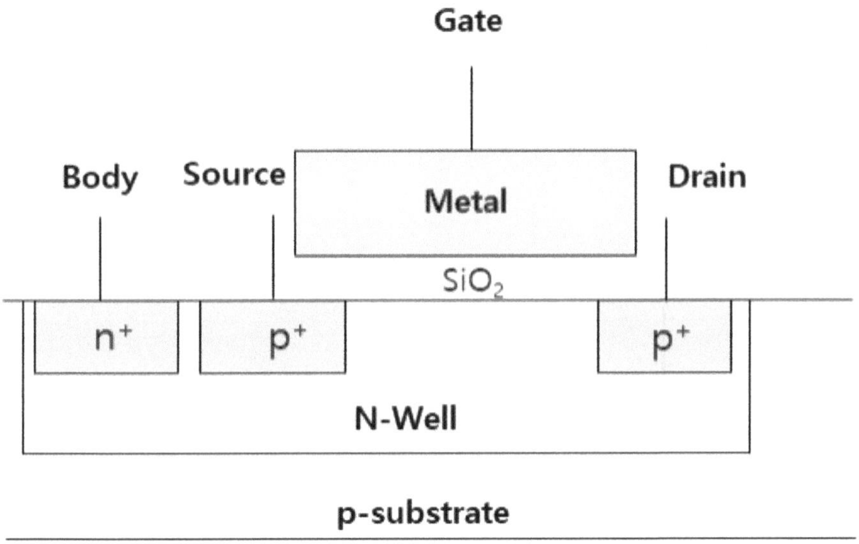


(b) process cross-sectional view

Fig. 4. PMOS transistor.

## 3.2 Current–Voltage Characteristics of NMOS Transistors

Fig. 5 shows the $I_{DS} - V_{DS}$ characteristic curve according to the $V_{GS}$ voltage of the NMOS transistor. The $I_{DS} - V_{DS}$ characteristic curve is a curve for the $I_{DS}$ current when the $V_{DS}$ voltage is sweeped while applying DC $V_{GS}$ voltages of 0 V, 1 V, 2 V, and 3 V.

The horizontal axis is the $V_{DS}$, and the vertical axis is the $I_{DS}$. If the $V_{GS}$ voltage is smaller than $V_T$, $I_{DS}$=0. So the $I_{DS}$ curve is parallel to the x-axis. When $V_{GS}$=1V, the $I_{DS}$ curve has curvature in the linear region and flat in the saturation as shown in Fig. 5. In the case of $V_{GS}$ larger than $V_T$, $I_{DS}$ increases according to the $V_{GS}$ voltage increases as shown in Fig. 5.

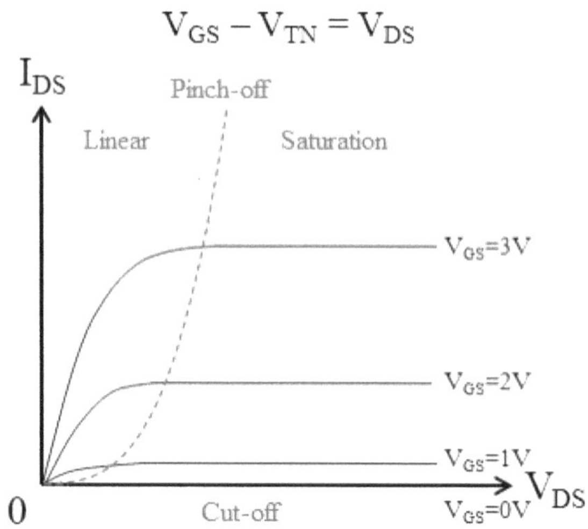


Fig. 5. $I_{DS}-V_{DS}$ curve of NMOS transistor.

Fig. 6(a) is a cross-sectional view of an NMOS transistor in a cut-off state. If the gate voltage $V_{GS}$ is increased from 0V to $V_{TN}$, the depletion region width with ionization of the acceptor impurities gradually increases on the silicon surface under the gate oxide. When the $V_{GS}$ voltage is applied to the $V_{TN}$, the depletion region stops increasing at the maximum depletion width and an inversion layer of electrons with a concentration higher than the acceptor impurity concentration of the p-substrate is formed. The inversion layer is formed between the n+ source region and the drain region.

When positive $V_{DS}$ voltage is applied, $I_{DS}$ current flows from the drain to the source because the electrons pass from the source to the drain through the inversion layer. In the NMOS transistor, the n-type inversion layer is called the n-channel because it is the channel through which the $I_{DS}$ current flows. The linear region can be seen when biased such that $V_{GS} \geq V_{TN}$ and $V_{GD} \geq V_{TN}$, and the n-channel is connected between the drain and the source, as shown in Fig. 6(b). The $I_{DS}$ equation is as shown in Eq. (2).

As the $V_{GS}$ voltage increases in the linear region, more electrons are induced in the n-channel, so the $I_{DS}$ current increases as shown in Fig. 5. If the $V_{DS}$ voltage is small and the $V_{DS}$ voltage increases with respect to the same $V_{GS}$ voltage, the $I_{DS}$ current increases linearly with $V_{DS}$.

In the pinch-off region shown in Fig. 6(c), when $V_{GS} \geq V_{TN}$ and $V_{GD} = V_{TN}$, the n-channel is barely maintained in the drain region, which is the boundary between the linear region and the saturation region.

If $V_{GS} \geq V_{TN}$ and $V_{GD} < V_{TN}$, the pinch-off point of the n-channel is shifted to the source region and the $I_{DS}$ is no longer increased by the $V_{DS}$ voltage as shown in Fig. 6(d). Eq. (3) is the $I_{DS}$ equation when the NMOS transistor is in the saturation region.

Next, $V_{GS}$ and $V_{GD}$ bias voltage conditions and $I_{DS}$ current equations for cut-off region, linear region and saturation region are summarized.

### (1) cut-off region: when $V_{GS} < V_{TN}$ and $V_{GD} < V_{TN}$

$$I_{DS} = 0 \tag{1}$$

### (2) linear region: when $V_{GS} \geq V_{TN}$ and $V_{GD} \geq V_{TN}$

$$I_{DS} = \mu_n Cox \frac{W}{L}\left\{(V_{GS} - V_{TN})V_{DS} - \frac{1}{2}V_{DS}^2\right\} \tag{2}$$

where $\mu_n$ is the electron mobility in $m^2/V \cdot s$ and $C_{ox}$ is the gate oxide capacitance per unit gate area in $F/m^2$.

**(3) saturation region: when $V_{GS} \geq V_{TN}$ and $V_{GD} < V_{TN}$**

$$I_{DS} = \frac{1}{2}\mu_n Cox \frac{W}{L}(V_{GS} - V_{TN})^2 \tag{3}$$

$C_{ox}$ is given by Eq. (4).

$$Cox = \frac{\epsilon_0 \epsilon_{ox}}{t_{ox}} \tag{4}$$

where $\epsilon_0$ (=$8.854\times10^{-12}$F/m) is the dielectric constant in vacuum and $\epsilon_{ox}$ is the relative dielectric constant of oxide having value of 3.9. The gate oxide thickness, $t_{ox}$, is determined by the process technology used to fabricate the MOS transistor.

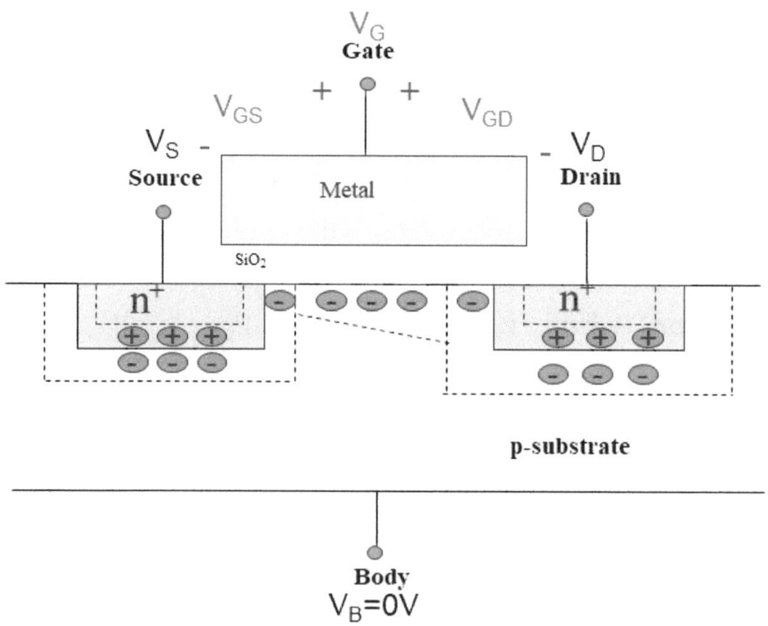


(a) cut-off region

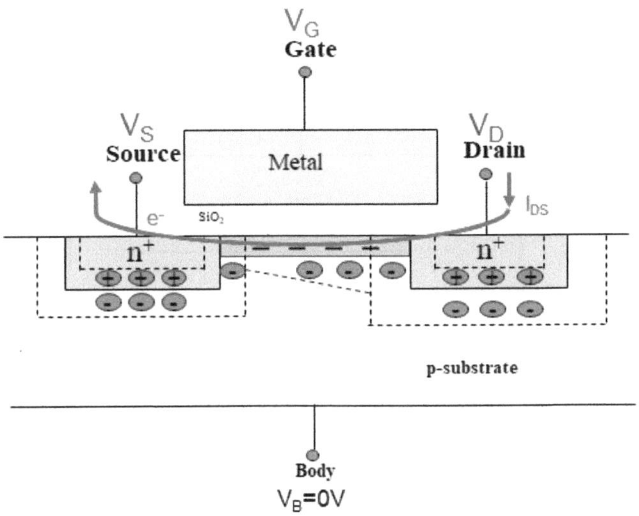


(b) linear region

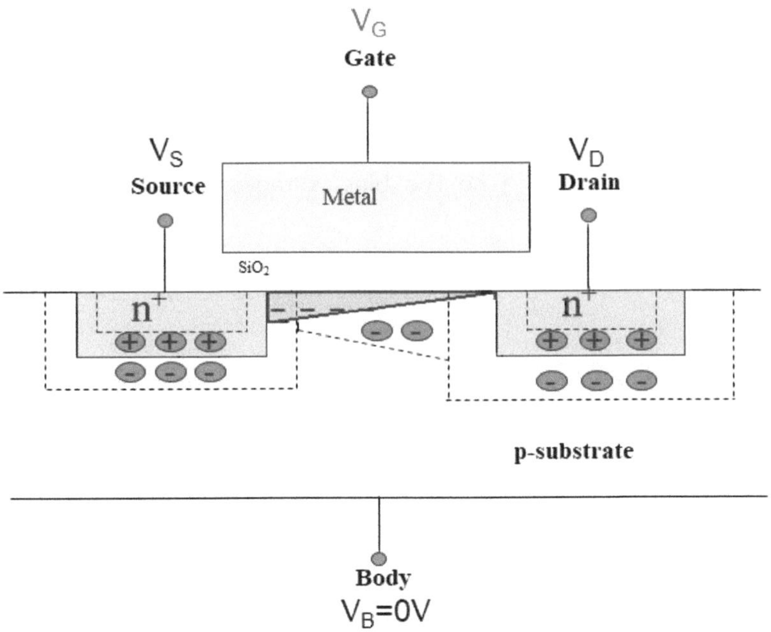


(c) pinch-off region

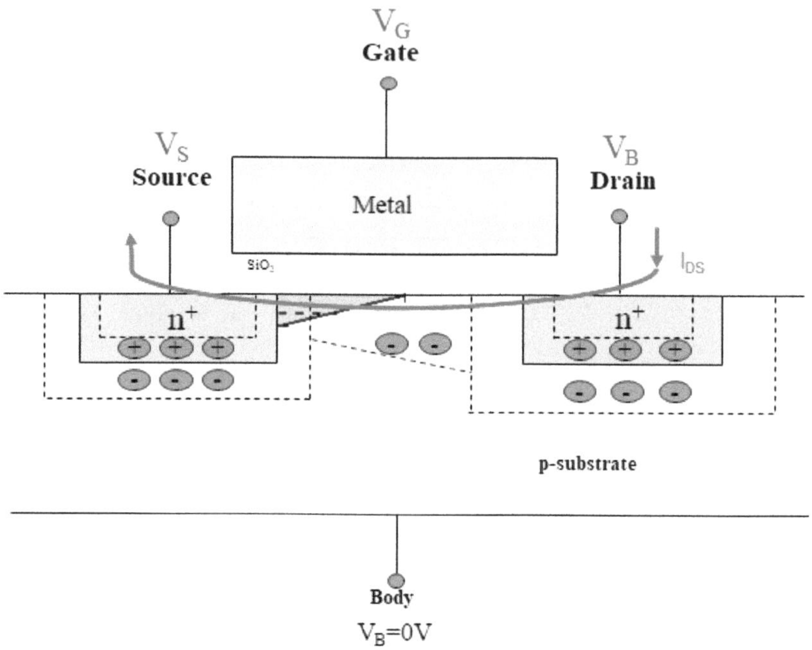


(d) saturation region

Fig. 6. The operating ranges of the NMOS transistor according to the bias voltages.

Example 3-1

*Calculate the $C_{ox}$ of a MOS transistor with a gate oxide thickness $t_{ox}$ of 100Å.*

$$Cox = \frac{\epsilon_0 \epsilon_{ox}}{t_{ox}} = \frac{3.9 \times 8.854 \times 10^{-12}}{100 \times 10^{-10}} = 3.45 \times 10^{-3}\ F/m^2 = 3.45 fF/\mu m^2 \quad (5)$$

**Example 3-2**

*Identify the operation region of the NMOS transistor and calculate the current I:* $\mu_n C_{ox}$ *= 70 μA/V², W=10 μm, L=0.25 μm, and* $V_{TN}$*=0.5 V.*

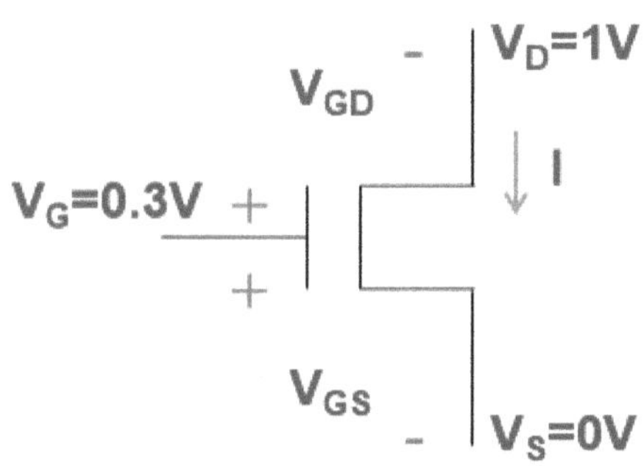


Fig. 7. $V_D = 1$ V, $V_G = 0.3$ V, and $V_S = 0$ V.

As $V_{GS}$ = 0.3 V, $V_{GS}$ is smaller than $V_{TN}$ and as $V_{GD}$ = −0.7 V, $V_{GD}$ is smaller than $V_{TN}$, which mean the cut-off region. The current $I_{DS}$ is 0 because the transistor is in the cut-off region.

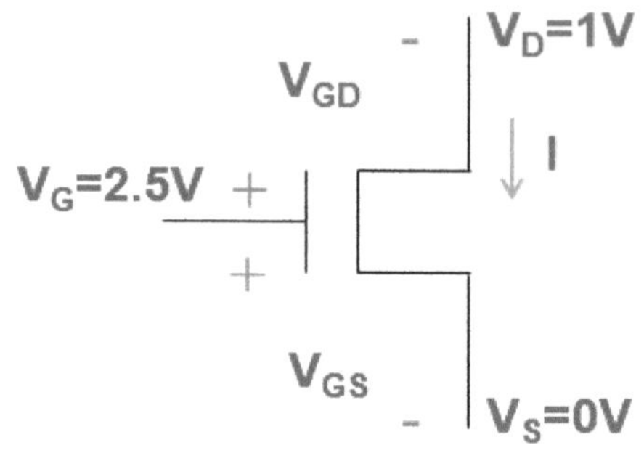


Fig. 8. $V_D = 1\ V$, $V_G = 2.5\ V$, and $V_S = 0\ V$.

As $V_{GS} = 2.5\ V$ , $V_{GS}$ is larger than $V_{TN}$ and as $V_{GD} = 1.5\ V$, $V_{GD}$ is larger than $V_{TN}$, so it corresponds to the linear region. The current $I_{DS}$ is 4.2 mA from the $I_{DS}$ equation for the linear region.

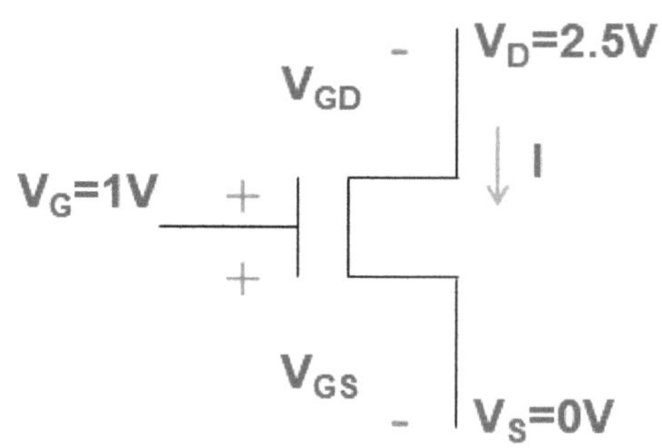


Fig. 9. $V_D = 2.5\,V$, $V_G = 1\,V$, and $V_S = 0\,V$.

As $V_{GS} = 1\ V$, $V_{GS}$ is larger than $V_{TN}$ and as $V_{GD} = -1.5\ V$ , $V_{GD}$ is smaller than $V_{TN}$, so it corresponds to the saturation region. The current $I_{DS}$ is 350 μA from the $I_{DS}$ equation for the saturation region.

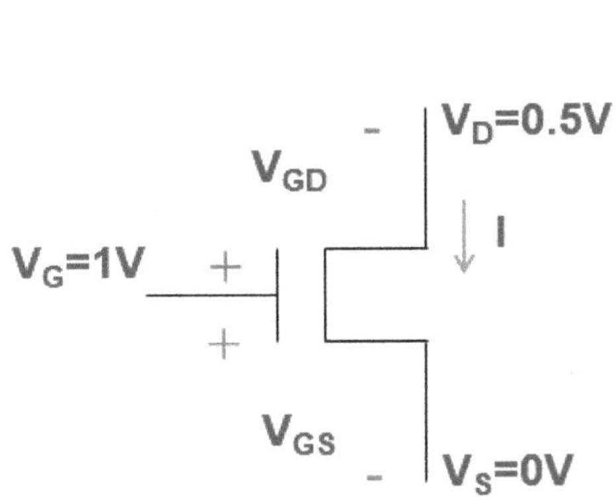


Fig. 10. $V_D = 0.5\ V,\ V_G = 1\ V$, and $V_S = 0\ V$.

As $V_{GS} = 1\ V$, $V_{GS}$ is greater than $V_{TN}$ and as $V_{GD} = 0.5\ V$, $V_{GD}$ is the same as $V_{TN}$, so it corresponds to the pinch-off region. As the current $I_{DS}$, it does not matter to use either $I_{DS}$ corresponding to the linear region or $I_{DS}$ corresponding to the saturation region, so the $I_{DS}$ current is 350 μA.

---

Figs. 11(a) and 11(b) show $I_{DS} - V_{DS}$ curves of 1.5 V NMOS and 5 V NMOS transistors, respectively. Fig. 7(a) shows the $I_{DS} - V_{DS}$ characteristic curve for a 1.5 V NMOS transistor. It plots the $I_{DS}$ current when the $V_{DS}$ voltage is sweeped while applying $V_{GS}$ voltage of 0.5 V, 1 V and 1.5 V. The on current of a 1.5 V NMOS transistor increases as the $V_{GS}$ voltage increases, as shown in Fig. 11(a), and is 4.69 mA at a typical NMOS model parameter, Temp. = 25℃ and $V_{GS}$ = $V_{DS}$ = 1.5 V. Fig. 11(b) shows the $I_{DS} - V_{DS}$ characteristic curve for a 5 V NMOS transistor. It plots the $I_{DS}$ current when sweeping the $V_{DS}$ voltage while applying the $V_{GS}$ voltage in the order of 1 V, 2 V, 3 V, 4 V and 5 V. The on current of the 5 V NMOS transistor increases as the $V_{GS}$ voltage increases, as shown in Fig. 11(b), and is 5.24 ㎃ at the typical NMOS model parameter, Temp. = 25℃ and $V_{GS}$ = $V_{DS}$ = 5 V.

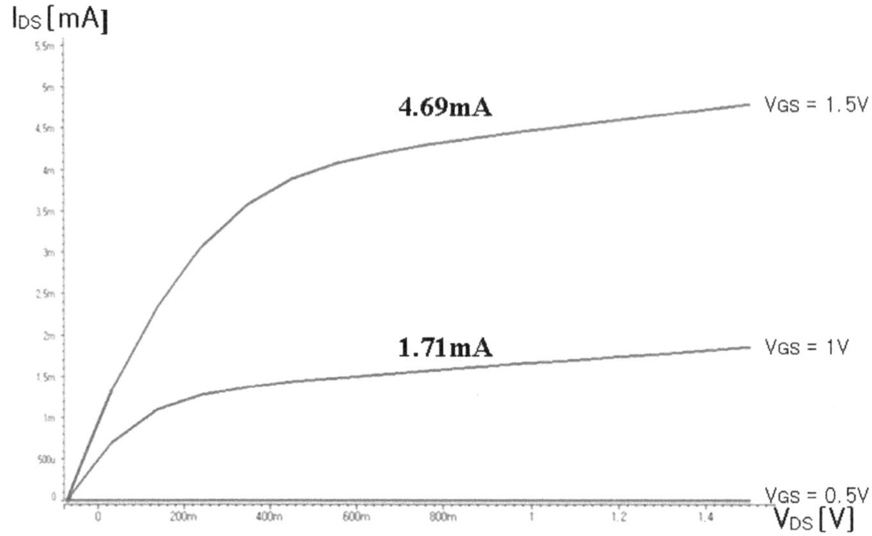


(a) 1.5V NMOS

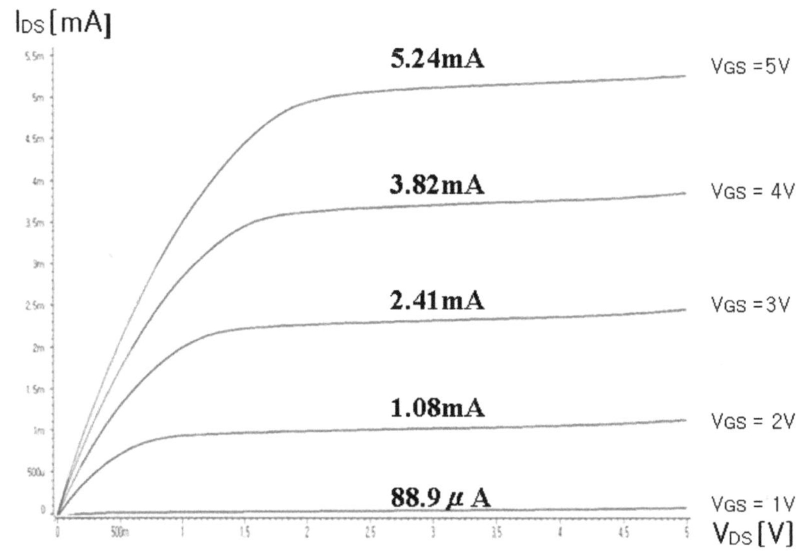


(b) 5V NMOS

Fig. 11. $I_{DS} - V_{DS}$ curves of NMOS transistors.

## 3.3 Current–Voltage Characteristics of PMOS Transistor

Fig. 12 shows the cross-sectional view of the operating region of the PMOS transistor according to the bias voltage. A PMOS transistor is formed on a N-Well in the p-substrate. First, N-Well is formed in the p-substrate. Next, on the surface of the N-Well, a $SiO_2$ layer, which is a gate oxide with good electrical characteristic, is formed, and then a metal gate is formed on the gate oxide. After the metal gate is formed, the p+ source and drain region are made by injecting acceptor impurities at a high concentration into a relatively low-doped N-Well. In Fig. 12, one of the p+ diffused regions is the source, the other is the drain, and the n+ diffused region formed in the N-Well is the body. The metal gate overlaps the p+ source and drain regions. A PMOS transistor is a 4-terminal device of source, drain, gate, and body, and no current flows from the drain to the source unless a p-channel is formed between the source and the drain.

As shown in Fig. 12, each depletion region at the source junction and drain junction are mainly formed in the N-Well which is the lightly doped region. Also, the depletion region width at the drain junction is wider than that at the source junction because the reverse bias voltage at the drain junction is larger than that at the source junction.

In the process sectional view of the PMOS transistor, the source and drain junctions are formed with parasitic pn junction diodes. In order for the MOS device to operate normally, a bias must be applied in the reverse direction to turn off these pn junction diodes. In many analog and digital integrated circuit designs, the N-Well voltage, the body of a PMOS transistor, is applied with a voltage VDD that is greater than the maximum voltage VDD applied to the source and drain nodes.

In the PMOS transistor, when the $V_{GS}$ voltage is 0 V, no channel is formed under the PMOS gate oxide, so the $I_{SD}$ current is 0 as shown in Eq. (6). Fig. 12(a) is a cross-sectional view of the PMOS transistor in the cut-off state. When a negative voltage is applied from 0V to $V_{TP}$ at a gate voltage of $V_{GS}$, the depletion region width at which the donor impurity is ionized is gradually increased on the silicon surface under the gate oxide.

When the $V_{GS}$ voltage is applied to the $V_{TP}$ voltage, the depletion region stops increasing at the maximum depletion width and an inversion layer is formed by holes having a concentration higher than that of the N-Well donor impurities. This inversion layer is formed between the p+ source region and the drain region. When negative $V_{DS}$ voltage is applied, $I_{SD}$ current flows from the source to the drain because the holes move from the source to the drain through the inversion layer.

In a PMOS transistor, a p-type inversion layer is called a p-channel because it is a path through which $I_{SD}$ current flows. The linear region can be seen when $V_{GS} \leq V_{TP}$ and $V_{GD} \leq V_{TP}$, and the p-channel is connected between the source and the drain as shown in Fig. 12(b). And the $I_{SD}$ formula is as shown in Eq. (7). As the $V_{GS}$ voltage in the linear region decreases to the negative, more holes are induced in the p-channel, resulting in an increase in the $I_{SD}$ current. Also, when the $V_{DS}$ voltage decreases to the negative with respect to the same $V_{GS}$ voltage, when the absolute value of the $V_{DS}$ voltage is small, the $I_{SD}$ current increases linearly with $V_{DS}$.

The p-channel is barely maintained in the drain region when the pinch-off region is seen when $V_{GS} \leq V_{TP}$ and $V_{GD} = V_{TP}$, which is the boundary between the linear region and the saturation region. If $V_{GS} \leq V_{TP}$ and $V_{GD} > V_{TP}$ are biased, the pinch-off point of the p-channel is shifted to the source region and the $I_{SD}$ does not increase any more with the $V_{DS}$ voltage as shown in Fig. 12(c). Eq. (8) is the $I_{SD}$ equation when the PMOS transistor is in the saturation region.

**(1) cut-off region: when $V_{GS} > V_{TP}$ and $V_{GD} > V_{TP}$**

$$I_{SD} = 0 \tag{6}$$

**(2) linear region: when $V_{GS} \leq V_{TP}$ and $V_{GD} \leq V_{TP}$**

$$I_{SD} = \mu_P Cox \frac{W}{L}\left\{(V_{GS} - V_{TP})V_{DS} - \frac{1}{2}V_{DS}^2\right\} \tag{7}$$

where $\mu$p is the hole mobility and the unit is ㎡/V·s.

**(3) saturation region: when $V_{GS} \leq V_{TP}$ and $V_{GD} > V_{TP}$**

$$I_{SD} = \frac{1}{2}\mu_p Cox \frac{W}{L}(V_{GS} - V_{TP})^2 \tag{8}$$

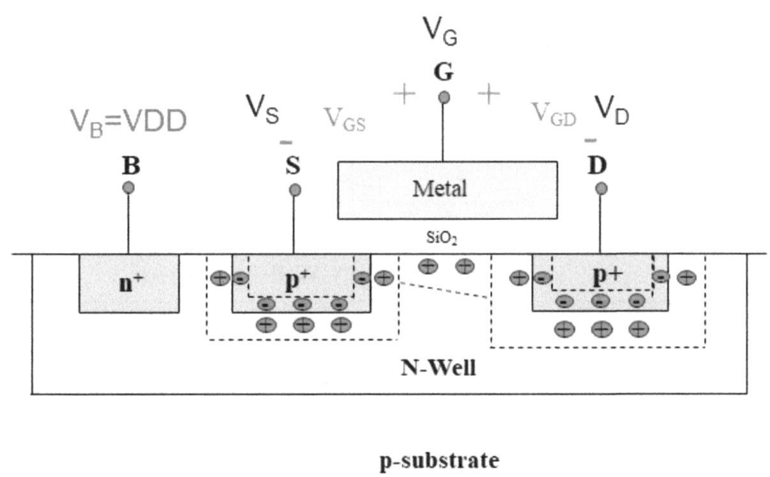


(a) cut-off region

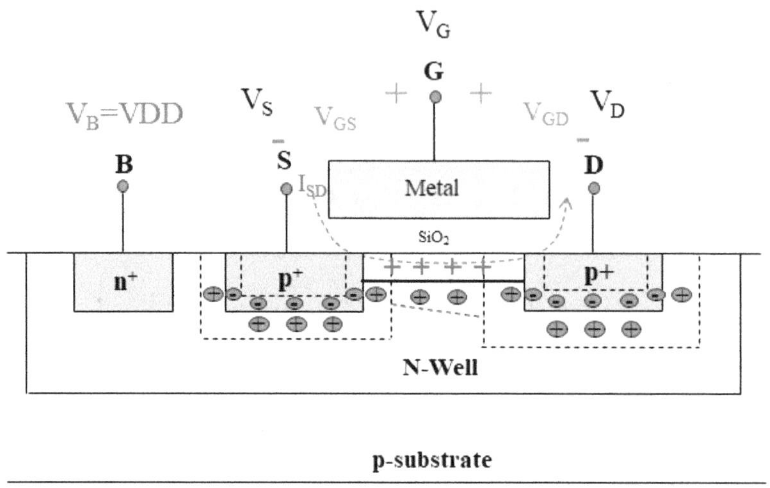


(b) linear region

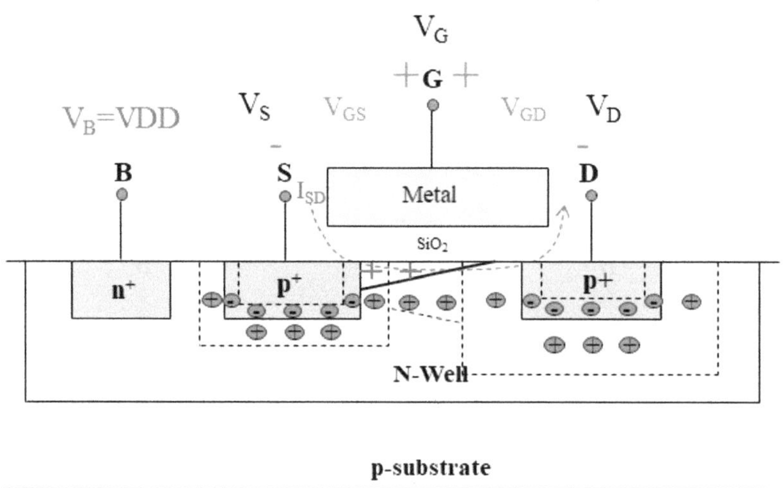


(c) saturation region

Fig. 12. Operating regions of PMOS transistor according to bias voltages.

## Example 3-3

*Identify the operating region of the PMOS transistor and Calculate the current I. Here,* $\mu_p C_{ox}$*=35 ㎂/V², $V_{TP}$=0.5 V, W=10 ㎛, and L=0.25 ㎛.*

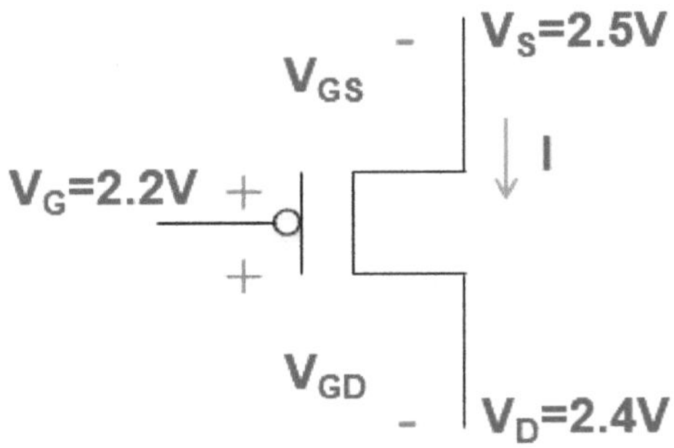


Fig. 13. $V_S$=2.5 V, $V_G$=2.2 V, and $V_D$=2.4 V.

As $V_{GS}$ = -0.3 V, $V_{GS}$ is greater than $V_{TP}$ and as $V_{GD}$ = -0.2 V, $V_{GD}$ is greater than $V_{TP}$, which mean the cut-off region. The current I (= $I_{SD}$) is 0 because the transistor is in the cut-off region.

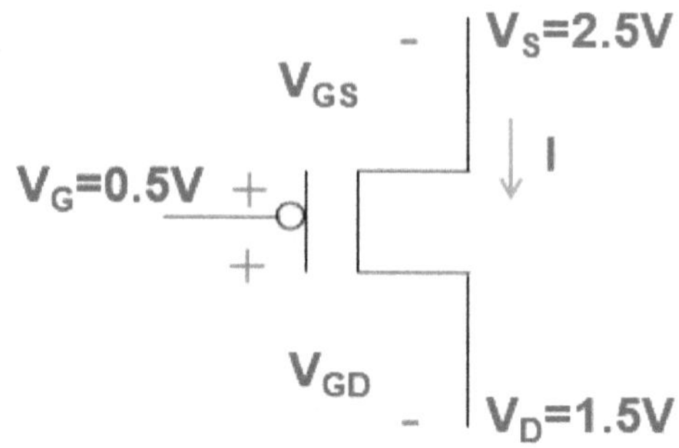


Fig. 14. $V_S$=2.5 V, $V_G$=0.5 V, and $V_D$=1.5 V.

As $V_{GS}$ = 2 V, $V_{GS}$ is larger than $V_{TP}$, and as $V_{GD}$ = −1 V, $V_{GD}$ is smaller than $V_{TP}$, so it corresponds to the linear region. The current I (= $I_{SD}$) is 1.4 mA from the $I_{SD}$ equation for the linear region.

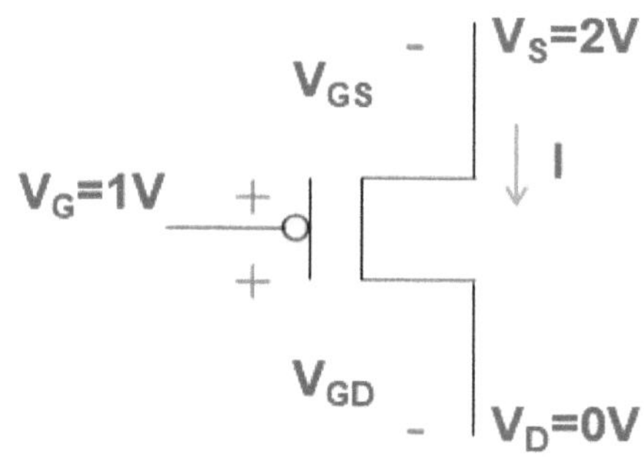


Fig. 15. $V_S$=2 V, $V_G$=1 V, and $V_D$=0 V.

As $V_{GS}$ = −1 V, $V_{GS}$ is smaller than $V_{TP}$ and as $V_{GD}$ = 1 V, $V_{GD}$ is larger than $V_{TP}$, so it corresponds to the saturation region. The current I (= $I_{SD}$) is 175 μA because it uses the $I_{SD}$ equation for the saturation region.

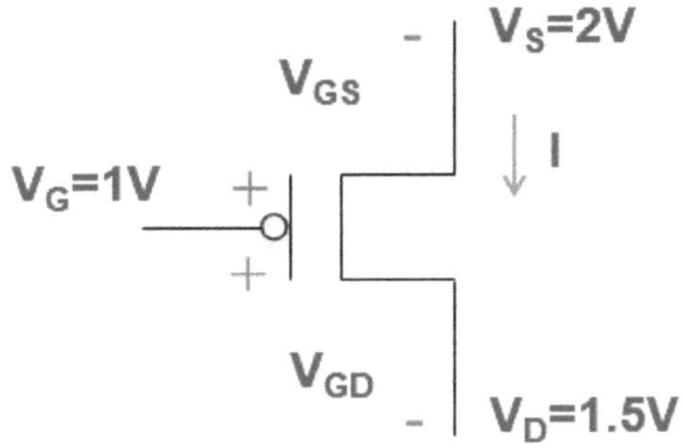


Fig. 16. $V_S$=2 V, $V_G$=1 V, and $V_D$=1.5 V.

As $V_{GS}$ = −1 V, $V_{GS}$ is smaller than $V_{TP}$ and as $V_{GD}$ = −0.5 V, $V_{GD}$ is the same as $V_{TP}$, so it corresponds to the pinch-off region. As the current, it does not matter to use either $I_{SD}$ corresponding to the linear region or $I_{SD}$ corresponding to the saturation region, so the I (= $I_{SD}$) current is 175 μA.

---

Figs. 17(a) and 17(b) show the $I_{SD} - V_{SD}$ curves of 1.5 V PMOS and 5 V PMOS transistors, respectively. Fig. 17(a) is the $I_{SD} - V_{SD}$ characteristic curve of 1.5 V PMOS transistor. It shows the curve of $I_{SD}$ current when sweeping $V_{SD}$ voltage while applying $V_{SG}$ voltage in the order of 0.5 V, 1 V, and 1.5 V. The on current of the 1.5 V PMOS transistor increases as the $V_{SG}$ voltage increases, as shown in Fig. 17(a), and is 2.5 mA at the typical PMOS model parameter, Temp. = 25℃ and $V_{SG}$ = $V_{SD}$ = 1.5 V.

Fig. 17(b) is the $I_{SD} - V_{SD}$ characteristic curve for 5 V PMOS transistor. It shows the curve of $I_{SD}$ current when sweeping $V_{SD}$ voltage while applying $V_{SG}$ voltage in the order 1 V, 2 V, 3 V, 4 V and 5 V. The on current of the 5 V PMOS transistor increases as the $V_{SG}$ voltage increases, as shown in Fig. 17(b), and is

3.05 mA at the typical PMOS model parameter, Temp. = 25℃ and $V_{SG}$ = $V_{SD}$ = 5 V. Since the electron mobility $\mu_n$ in the MOS device is larger than the hole mobility $\mu_p$, the NMOS transistor has a larger current I than the PMOS transistor for the same W/L in the I-V curve simulation of the NMOS and PMOS transistors.

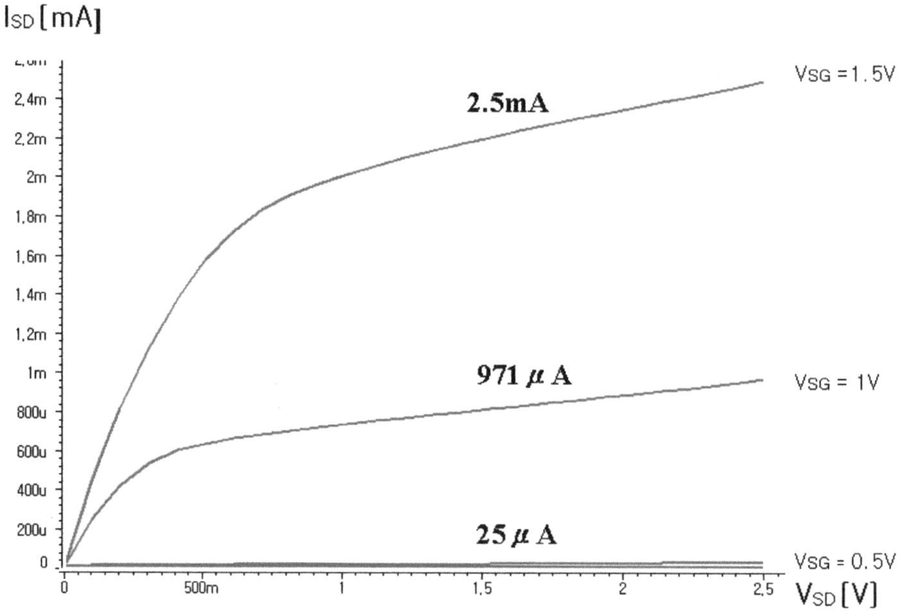


(a) 1.5 V PMOS

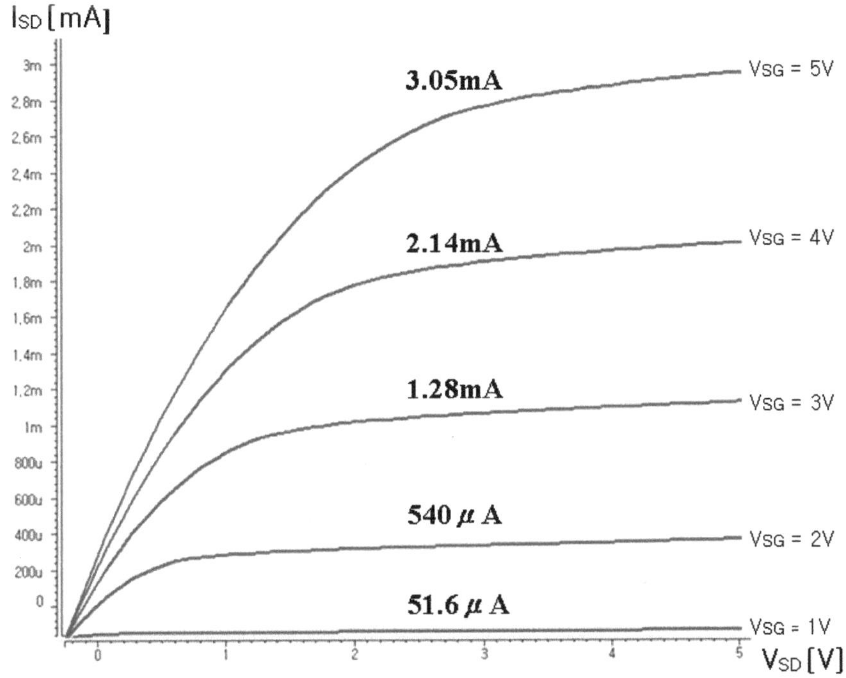


(b) 5V PMOS

Fig. 17. I-V curves of PMOS transistors.

## 3.4 Threshold Voltage

Fig. 18(a) shows a process cross-sectional view of an ideal MOS capacitor in an equilibrium state with an applied voltage V = 0 V, and Fig. 18(b) shows the energy band diagram of an ideal MOS capacitor case.

The work function of the metal is the energy required to move one electron out of the metal from the Fermi level. On the other hand, in MOS devices, it is more convenient to use the modified work function $q\Phi_m$, which is a measurement from the metal Fermi level to the conduction band of the oxide for the metal-oxide interface. Similarly, $q\Phi_s$ is the modified work function in the semiconductor-oxide interface.

In the ideal MOS capacitor, it is assumed that there is no difference between two modified work functions, so $q\Phi_m = q\Phi_s$. And $q\phi_F$ in Fig. 18(b) is the measurement of the position of $E_F$ below $E_i$, indicating how strongly the semiconductor is p-type.

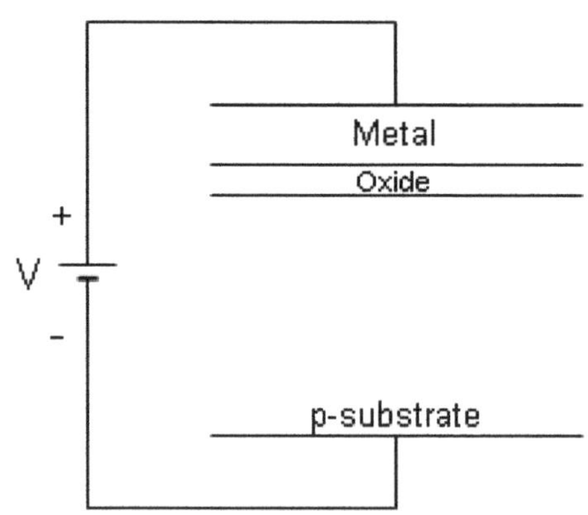


(a) process cross-sectional view

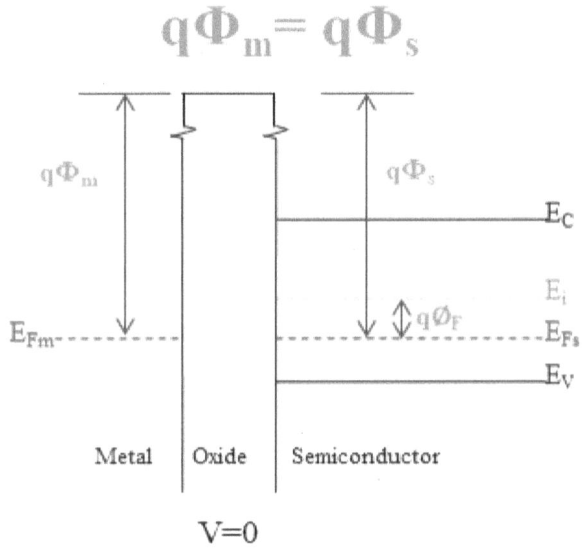


(b) energy band diagram

Fig. 18. Process cross-sectional view and energy band diagram for ideal MOS capacitor in equilibrium state.

Since $\Phi_m$ and $\Phi_s$ do not change with applied voltage, the fact that $E_{Fm}$ is lower than $E_{Fs}$ by qV as shown in Fig. 19(b) causes tilt in the conduction band of oxide. The electric field E(x) gives a slope in $E_i$, $E_v$ and $E_c$ as shown in Eq. (9).

$$E(x) = \frac{1}{q}\frac{dE_i}{dx} \tag{9}$$

Fig. 19 shows the process cross-sectional view and energy band diagram in depletion mode where $0\,V < V < V_{TN}$ and $\phi_F > \phi_S$ in an ideal MOS capacitor. The potential of the metal gate rises and $E_{Fm}$ goes down by qV from its position in equilibrium. As a result, the oxide conduction band is tilted as shown in Fig. 19(b). The positive voltage applied to the metal adheres positive charges to the metal and generates negative charges that match the surface of the semiconductor. In the case of a p-substrate, holes from the semiconductor surface are depleted as shown in Fig. 19(a), leaving negatively charged ionized acceptors. As shown in Fig. 19(b), $E_i$ moves closer to $E_F$ and the energy band bends down near the semiconductor surface.

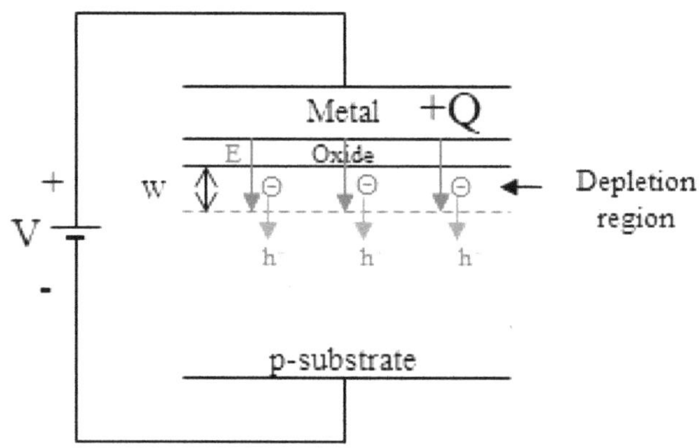


(a) process cross-sectional view

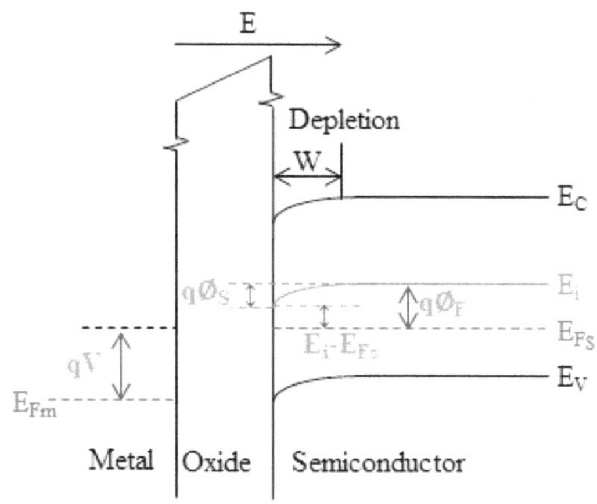


(b) energy band diagram

Fig. 19. Process cross-section and energy band diagram for an ideal MOS capacitor in a depletion state.

Eq. (10) is for the hole concentration of the semiconductor surface.

$$p = n_i e^{(E_i - E_F)/kT} \tag{10}$$

As the $E_i - E_F$ at the semiconductor surface is reduced, the hole concentration will be reduced and then negatively charged ionized acceptors will be maintained.

The inversion region in Fig. 20 can be divided into the weak inversion and the strong inversion. Weak inversion is the case where $0\,V < V < V_{TN}$ and $\phi_F < \phi_S < 2\phi_F$, and strong inversion is $V \geq V_{TN}$ and $\phi_S = 2\phi_F$. In the inversion of Fig. 20(b), if we continue to increase the positive voltage V in the metal, $E_i$ will bend below $E_{Fs}$. The fact that $E_{Fs} - E_i$ is positive means that it has a large electron concentration in the conduction band as shown in Eq. (11).

$$n = n_i e^{(E_F - E_i)/kT} \tag{11}$$

This inversion layer is not from the impurity doping but from the inversion of the p-type semiconductor to the n-type by the voltage V applied to the metal gate. This inversion layer is separated from the p-type semiconductor by the depletion region. Weak inversion means that the electron concentration in the inversion layer is lower than the hole concentration, which is the majority carrier created by the acceptor impurities in bulk. On the other hand, the strong inversion means that $\phi_s$ becomes $2\phi_F$ as shown in Eq. (12).

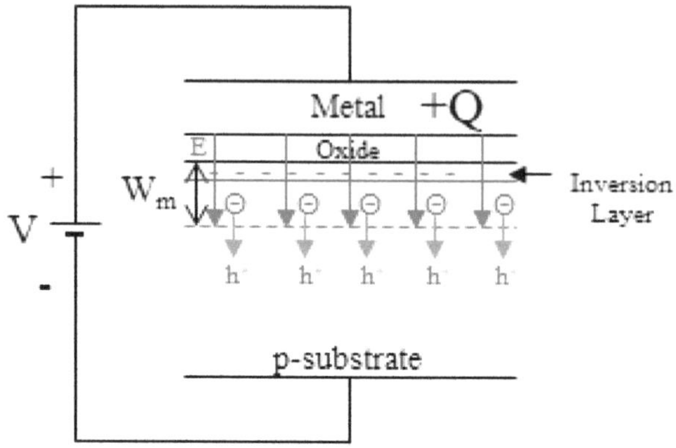


(a) process cross-sectional view

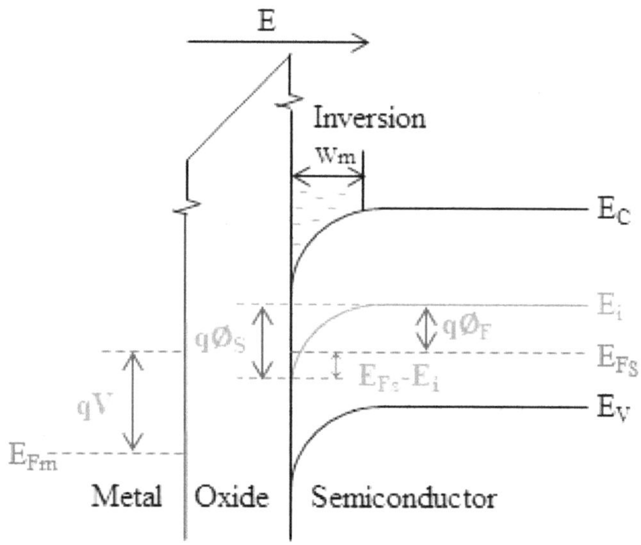


(b) energy band diagram

Fig. 20. Process cross-section and energy band diagram for an ideal MOS capacitor in the inversion state.

$$\phi_s(strong\ inversion) = 2\phi_F = 2\frac{kT}{q}ln\frac{N_a}{n_i} \tag{12}$$

In Fig. 20 (a), depletion width W is the same as in Eq. (13), and depletion width increases with increasing surface potential $\phi_s$ until strong inversion is achieved.

$$W = \sqrt{\frac{2\epsilon_0\epsilon_{Si}\phi_s}{qN_a}} \tag{13}$$

When $\phi_S = 2\phi_F$, which is the strong inversion, the depletion width becomes the maximum depletion width $W_m$ in Eq. (14).

$$W_m = \sqrt{\frac{2\epsilon_0\epsilon_{Si}\phi_s(strong\ inversion)}{qN_a}} = \sqrt{\frac{2\epsilon_0\epsilon_{Si}\times 2\phi_F}{qN_a}} = 2\sqrt{\frac{\epsilon_0\epsilon_{Si}kT\ln\frac{N_a}{n_i}}{q^2N_a}} \tag{14}$$

In the strong inversion, the depletion charge $Q_{d.\max}$ per unit area of the depletion region is given by Eq. (15).

$$Q_{d.\max} = -qN_aW_m = -2\sqrt{\epsilon_0\epsilon_{Si}qN_a\phi_F} \tag{15}$$

For an ideal MOS capacitor, the applied voltage V should be the sum of the voltage for making the depletion region charge $Q_{d.\max}$ and the surface potential $\phi_s$ (strong inversion) of $2\phi_F$ in Eq. (15). Therefore, in the case of an ideal MOS capacitor, a strong inversion occurs. The threshold voltage required is given by Eq. (16).

$$V_T = -\frac{Q_{d.\max}}{C_{ox}} + 2\phi_F \tag{16}$$

The ideal MOS device is assumed to be $q\Phi_m = q\Phi_s$, and it is assumed that electric charges do not exist in the insulator and there is no crystal defect in the semiconductor-substrate interface. However, the actual MOS device has a difference in work function between the ploy-silicon gate and p-substrate, electric charges exist in the insulator, and crystal defects exist in the semiconductor-oxide interface, so actual MOS devices are affected by the differences aforementioned differently from the ideal MOS devices. The flat band voltage $V_{FB}$ of Eq. (17) must be applied in order to make the band bended at the semiconductor surface due to the difference in work function and the interface charge to a flat band.

$$V_{FB} = \phi_{ms} - \frac{Q_i}{C_{ox}} \tag{17}$$

The $V_T$ of the actual MOS device can be obtained by adding the $V_{FB}$ of Eq. (17) to the $V_T$ obtained for the ideal MOS structure of Eq. (16). Eq. (18) corresponds to the $V_T$ equation of a practical MOS device, and corresponds to the case of $V_{SB}$ = 0 V without a body effect in a MOS transistor.

$$V_T = \phi_{ms} - \frac{Q_i}{C_{ox}} - \frac{Q_{d.\max}}{C_{ox}} + 2\phi_F \tag{18}$$

Fig. 21(a) is a cross-sectional view of the case where $V_{SB}$ = 0 V without MOS transistor body effect. The shaded region in Fig. 21 (a) corresponds to the depletion region charge formed under the gate oxide in the source region. If a positive voltage is applied to the source voltage $V_S$, the width of the depletion region increases as shown in Fig. 21(b) as the reverse bias voltage of the source junction increases.

Fig. 21(a) is a cross-sectional view of the case where $V_{SB}$ = 0 V without MOS transistor body effect. The shaded region in Fig. 21 (a) corresponds to the depletion region charge formed under the gate oxide in the source region. If a positive voltage is applied to the source voltage $V_S$, the width of the depletion region increases as shown in Fig. 21(b) as the reverse bias voltage of the source junction increases. Thus, as the $V_{SB}$ voltage increases, the depletion region charge under the gate oxide of the source region is increased by the body effect as shown in Fig. 21(b), and the threshold voltage is increased by $\triangle V_T$ similarly as shown in Eq. (19). In Eq. (19), $V_{T0}$ is the threshold voltage at $V_{SB}$ = 0 V, $\phi_F$ is the Fermi potential, and $\gamma$ is the body effect parameter given by Eq. (20).

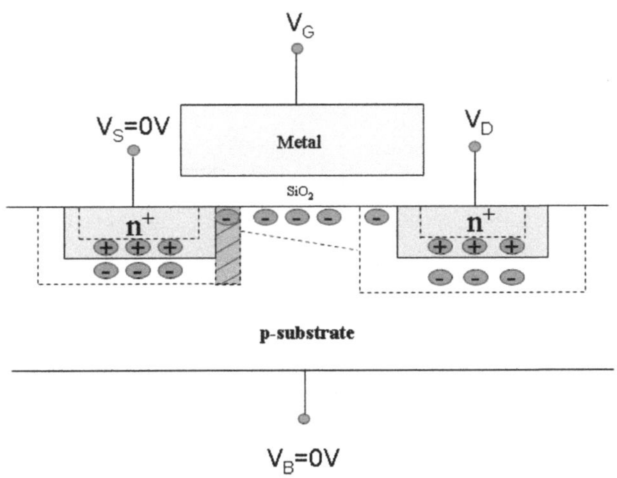


(a) without a body effect

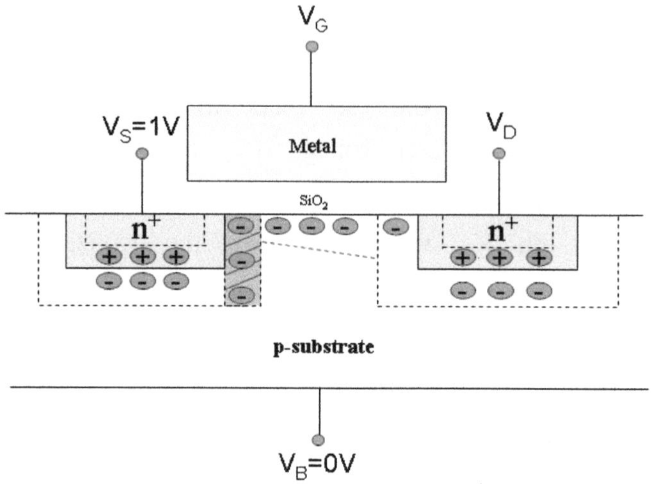


(b) with a body effect

Fig. 21. Process cross-sectional view of NMOS transistor.

$$V_T = V_{T0} + \Delta V_T = V_{T0} + \gamma\left(\sqrt{2\phi_F + V_{SB}} - \sqrt{2\phi_F}\right) \tag{19}$$

$$\gamma = \frac{\sqrt{2q\epsilon_0\epsilon_{Si}N_a}}{C_{ox}} \tag{20}$$

## 3.5 Channel Length Modulation Effect

When the NMOS transistor operates in the saturation region, the $I_{DS}$ current is constant regardless of the $V_{DS}$ voltage as shown in Fig. 22(a). As shown in Eq. (3), the $I_{DS}$ current has nothing to do with the $V_{DS}$ voltage because eq. (3) does not include the $V_{DS}$ voltage.

However, as the $V_{DS}$ voltage increases in the saturation region, the $I_{DS}$ current increases linearly as shown in Fig. 22(b). Fig. 22(a) shows the $I_{DS} - V_{DS}$ curve of the NMOS transistor without considering the channel length modulation effect, and Fig. 22(b) shows the case considering the channel length modulation effect. The $I_{DS}$ current increases linearly in the saturation region as shown in Fig. 22(b). The reason why the $I_{DS}$ current increases linearly in the saturation region is as follows. As the $V_{DS}$ voltage increases in the saturation region, the reverse bias voltage between the drain node and the body node increases. As the reverse bias voltage of the drain junction increases, the depletion region width of the drain junction region increases and the pinch-off at the end of the n-channel shifts from the drain to the source.

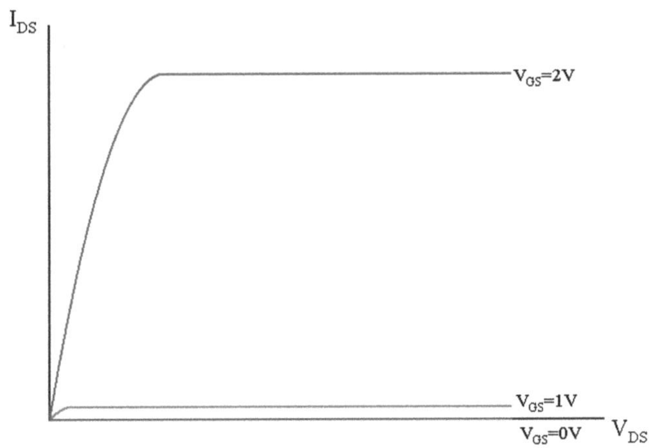


(a) the channel length modulation effect is considered

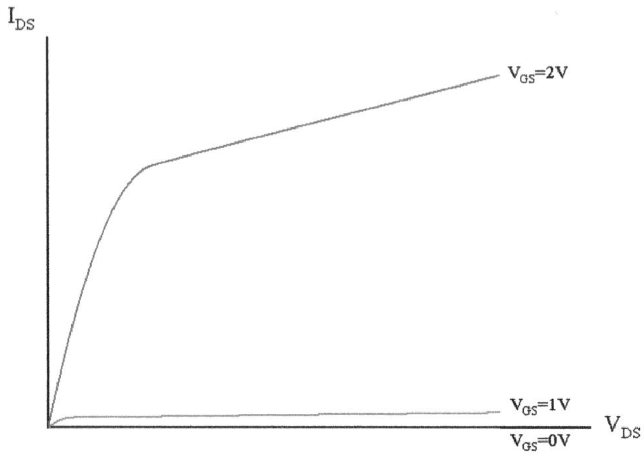


(b) the channel length modulation effect is not considered

Fig. 22. $I_{DS} - V_{DS}$ curve of NMOS transistor.

Fig. 23 shows a process cross-sectional view of the NMOS transistor in the saturation region. Figs. 23(a) and 23(b) show the cases when the $V_{DS}$ voltage is 1 V and 2 V, respectively, and the effective channel length decreases when the $V_{DS}$ voltage increases. The channel length modulation effect means that the $I_{DS}$ current increases linearly as the $V_{DS}$ voltage increases and effective channel length decreases. Fig. 23(b) shows the case when the $V_{DS}$ voltage is 2 V. Effective channel length for 2 V is shorter than that for 1 V.

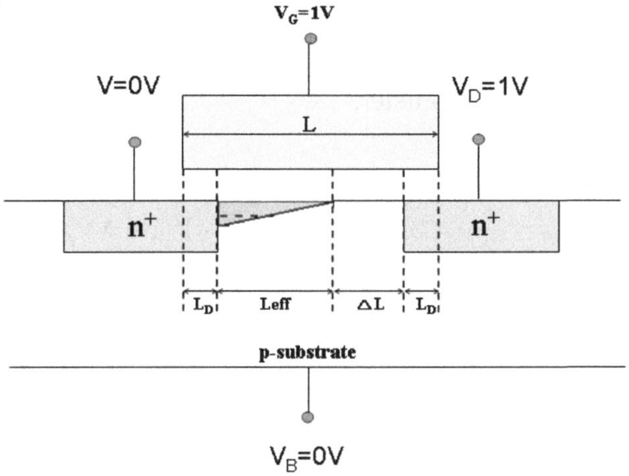


(a) in case of $V_{DS}$=1 V

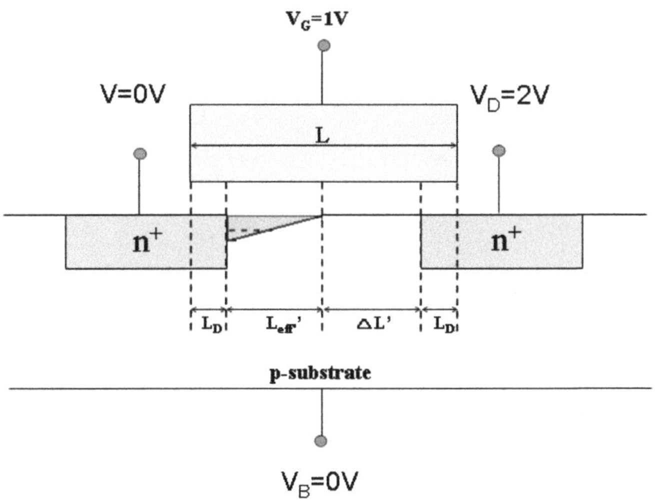


(b) in case of $V_{DS}$=2 V.

Fig. 23. Effective channel length variations according to $V_{DS}$ voltages in saturation region .

The $I_{DS}$ current considering the channel length modulation effect when the NMOS transistor is in the saturation region is shown in Eq. (21), where $\lambda$ is the channel length modulation parameter.

$$I_{DS} = \frac{1}{2}\mu_n Cox\frac{W}{L}(V_{GS} - V_{TN})^2(1 + \lambda V_{DS}) \tag{21}$$

## Exercises

1. Answer the following questions.

(a) Draw the symbol and process cross-sectional view of the NMOS transistor.

(b) Draw the symbol and process cross-sectional view of the PMOS transistor.

(c) When gate oxide thickness $t_{ox}$=70Å, calculate gate oxide capacitance $C_{ox}$ per μm$^2$. The unit of $C_{ox}$ is expressed as fF/μm$^2$.

2. Calculate the current I of the MOS transistor and distinguish the operating region. Here, $\mu_n c_{ox}$= 70 μA/V$^2$, $V_{TN}$=0.5 V, $\mu_p c_{ox}$= 35 μA/V$^2$, and $V_{TP}$=−0.5 V. W and L are marked on the circuit.

(a)

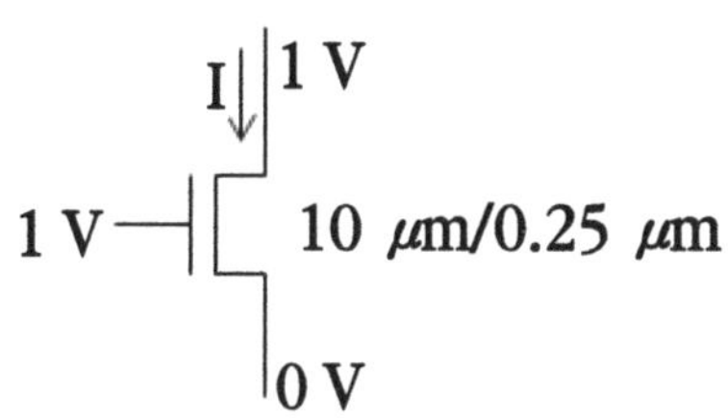

(b)

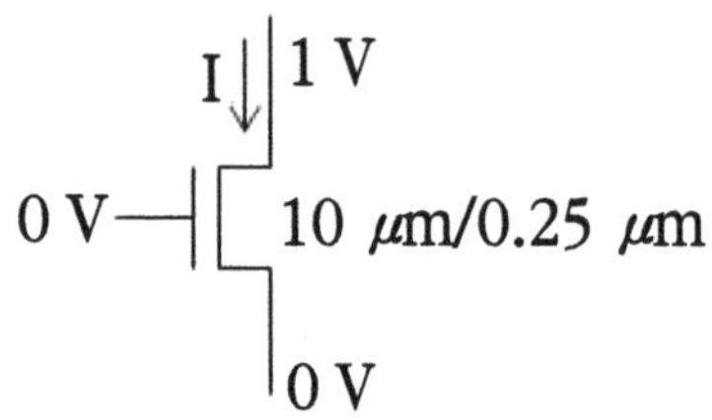


(c)

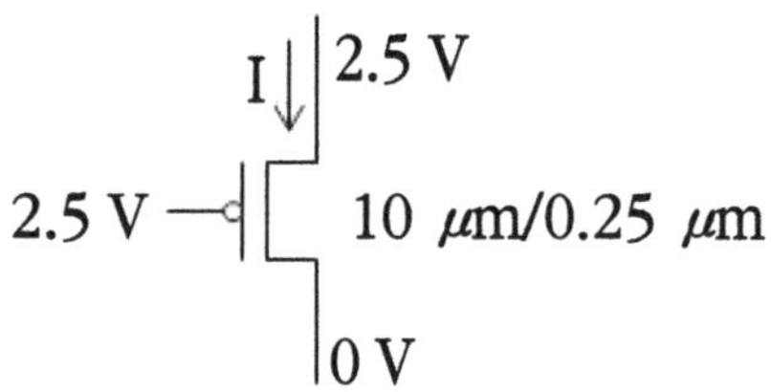


(d)

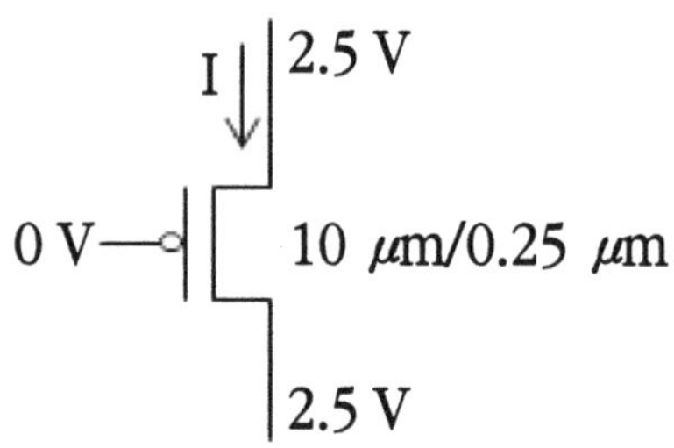


3. Draw the $I_C - V_{CE}$ curve of npn BJT.

4. Draw an energy band diagram for an ideal MOS capacitor.

(a) equilibrium state

(b) strong inversion

5. Answer the following questions.

(a) Calculate $\phi_S$ where MOS capacitor is inverted strongly for p-substrate of $N_a = 10^{15}$ acceptors/㎤ at T=300°K.

(b) Write the threshold voltage formula of the ideal MOS capacitor.

(c) Calculate the $V_T$ considering the body effect of NMOS transistor when $V_{T0}$=0.5 V, $2\phi_F$=0.8 V, $\gamma$=0.5, and $V_{SB}$=1 V.

6. Calculate the current I considering the channel length modulation effect of the circuit below. Here, $\mu_n C_{ox}$=70 μA/V$^2$, $V_{TN}$=0.5 V, W=10 μm, L=2 μm, and $\lambda$=0.02[V$^{-1}$].

(a)

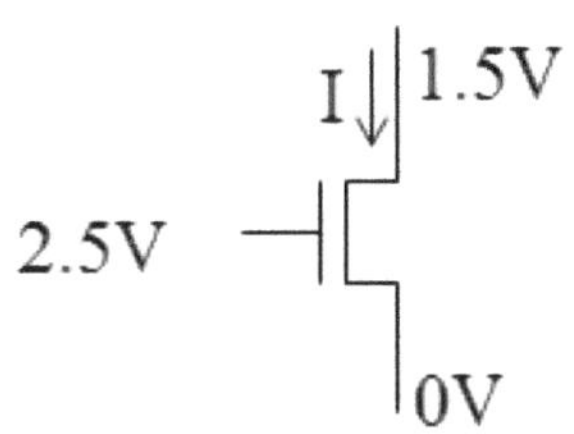

(b)

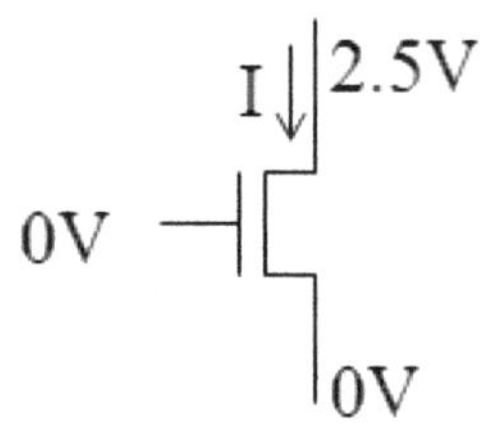


7. Calculate the current I of PMOS transistor. Here, $\mu_p c_{ox}$=35 μA/V$^2$, $V_{TP}$=−0.5 V, W=10 μm, and L=2 μm.

(a)

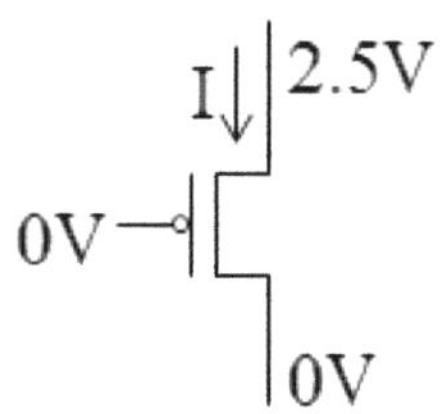


(b)

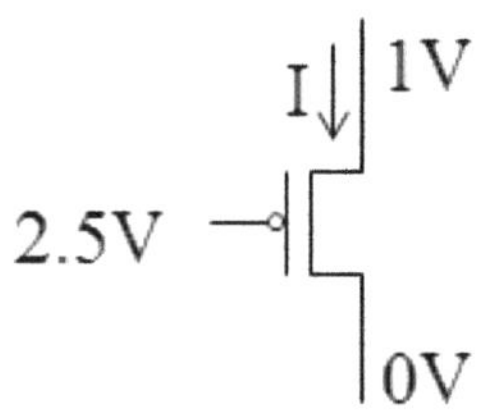

8. Draw the energy band diagram along the x-x' axis in equilibrium.

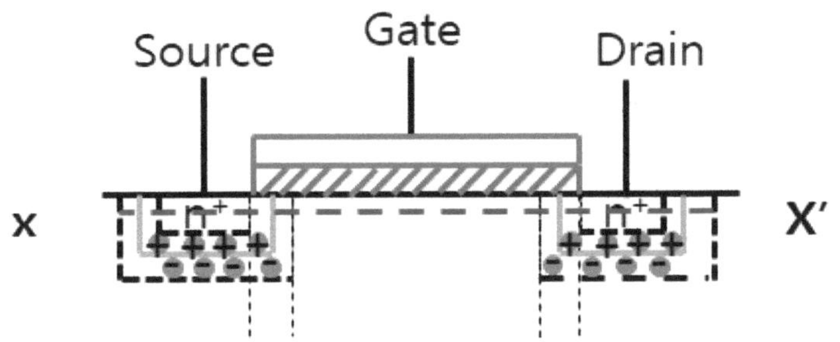


9. Answer the following questions.

(a) When VDD = 1.8 V, how many volts should the threshold voltage of the PMOS transistor be set to, considering $\frac{V_T}{VDD}$?

(b) For gate oxide thickness $t_{ox}$ = 100Å, calculate the gate oxide capacitor $C_{ox}$ per μm$^2$. Units should be expressed in fF/μm$^2$.

(c) Calculate the gate capacitance $C_G$ of the NMOS transistor with W = 1 μm and L = 1 μm in relation to the problem in (b).

(d) If VDD = 1.5 V, how many volts should be set for the $V_{TP}$ of the PMOS transistor, considering $\frac{V_{TP}}{VDD}$?

(e) For gate oxide thickness $t_{ox}$ = 50Å, calculate the gate capacitance $C_G$ of the NMOS transistor with W = 10 μm and L = 10 μm.

(f) Draw the $I_{DS}-V_{DS}$ curve of the NMOS transistor.

(g) What is the pinch-off voltage of the biased NMOS transistor as shown below? Here, $V_{TN}$ = 0.5 V.

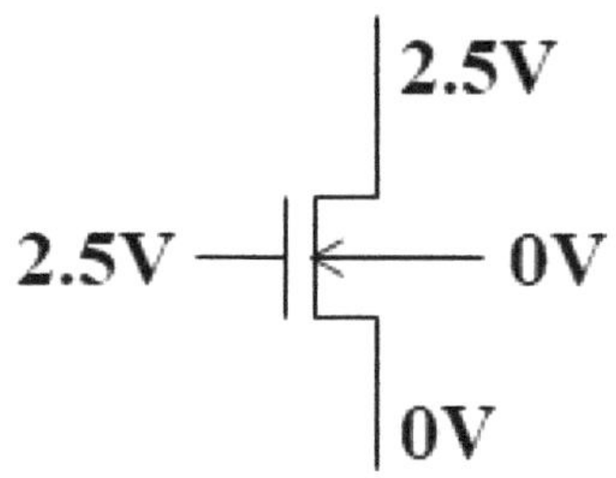


(h) What is the body voltage of NMOS transistor in digital IC design? Here, VDD = 2.5 V.

(i) Draw the $I_{DS}-V_{DS}$ curve of the NMOS transistor and mark the cut-off region, linear region, saturation region, and pinch-off region.

(j) What is the $V_{TN}$ target voltage for NMOS transistor of VDD = 1.5 V?

## Chap. 3 References

[1] Gyedal Kwak et al., Solid-State Electronics, Pearson, 2009.

[2] Kyungwook Shin, Design of CMOS Digital Integrated Circuits, Hanbit Academy, 2014.

[3] Seunghun Lee et al., Design of CMOS Analog/Mixed-Mode Integrated System, $\Sigma$ Sigma Press, 2004.

# 4 CHAPTER

# CMOS Process

## 4.1 CMOS Process

CMOS (Complementary Metal-Oxide-Semiconductor) process uses complementary symmetry of NMOS transistor and PMOS transistor. The CMOS logic process uses a twin-Well CMOS process using both N-Well and P-Well as shown in Fig. 1. The NMOS transistor is formed in the P-Well, and the PMOS transistor is formed in the N-Well. In the twin-Well CMOS process, the threshold voltage of the PMOS transistor and the NMOS transistor can be independently controlled using N-Well and P-Well.

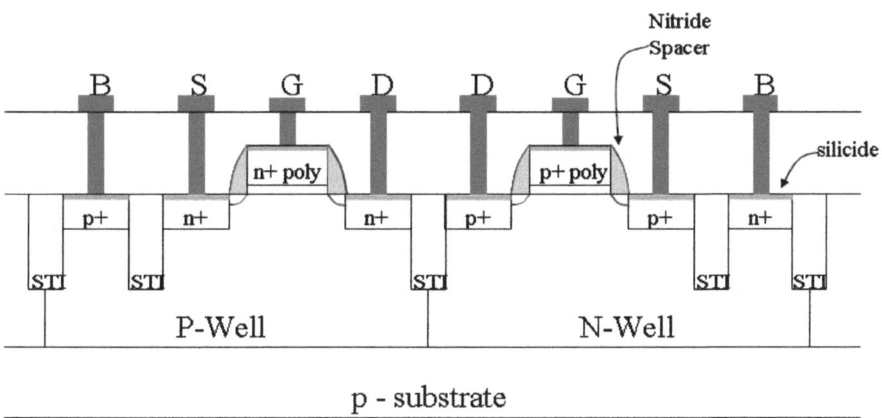


Fig. 1. Process cross-sectional view of twin-Well CMOS.

Semiconductor IC manufacturing process transfers the layout pattern of the circuit drawn on the mask to the silicon wafer of p-substrate using unit process technology such as oxidation, etching, ion implantation, diffusion, and deposition. The wafers that have been processed through the manufacturing process are packaged for the dies that pass the wafer test to make chips.

Semiconductor masks are made from layouts, and geometric patterns of the microstructures to be formed on the wafers are drawn on the semiconductor masks. Semiconductor layout is divided into full custom layout and auto P&R (Place and Route) depending on the method. Full custom layout is used mainly for layout of analog integrated circuits. Engineers place devices such as transistors, resistors, and capacitors used in design circuits and connect them with wiring manually according to design rules provided by a process. On the other, auto P&R is a method to place designed circuits at desired positions using library cells and to route them automatically, which is often used for laying out a digital integrated circuit as compared with a full custom layout.

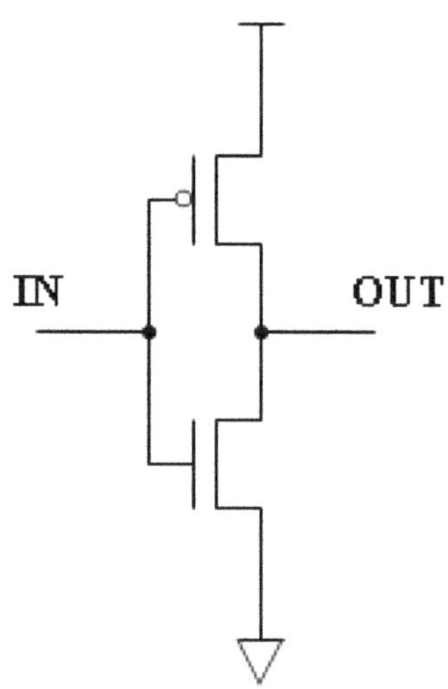


Fig. 2. Circuit of CMOS inverter.

The layout of the CMOS inverter circuit in Fig. 2 is shown in Fig. 3. The CMOS inverter circuit consists of PMOS and NMOS transistors. As shown in Fig. 2, the source of the PMOS transistor is connected to the supply voltage VDD, and the source of the NMOS transistor is connected to VSS (Virtual Ground). The gates of the NMOS and PMOS transistors are connected to the input signal IN, and the drains of the NMOS and PMOS transistors are connected to the output signal OUT. On the other hand, the N-Well which is the body of the PMOS transistor is connected to the VDD, and the P-Well which is the body of the NMOS transistor is connected to the VSS.

The layout images of the inverter using the twin-Well CMOS process are shown in Fig. 3. Fig. 3(a) shows the case where the top metal is M1, and Fig. 3(b) shows the layout image when the top metal is M2. In actual layout, layout can be done even if top metal is M1 as shown in Fig. 3(a), but a layout image is also added for the case where top metal is M2 for explanation of process section. The layers used in the layout are NWELL (N-Well), NACT (n-active), PACT (p-active), NPSD (n+ S/D implant), PPSD (p+ S/D implant), POLY, M1 (metal 1), VIA1 (via 1), and M2 (metal 2) are used. In the twin-Well CMOS process where the top metal is M1, both the N-Well and the P-Well process masks are required. Therefore, the PWELL (P-Well) not shown in Fig. 3(a) is created using the non-NWELL region when mask tooling. The PMOS transistor of Fig. 3(a) is formed at the intersection of the PACT region inside the NWELL with the gate POLY. Since the M1 layer overlaps the CONT layer, the p+ diffusion region which is the source of the PMOS transistor and the n+ region which is the body are applied with the VDD voltage. On the other hand, the NMOS transistor is formed at the intersection of the gate POLY with the NACT region inside the PWELL which is not the NWELL. Since the M1 layer overlaps the CONT layer, the n+ diffusion region which is the source of the NMOS transistor and the p+ diffusion region which is the body are applied with the VSS voltage. The IN signal, M1, is connected to the gate POLY of the NMOS and PMOS transistors using the CONT layer. The n+ diffusion, which is the drain of the NMOS transistor, and the p+ diffusion, which is the drain of the PMOS transistor, can output to the OUT signal through M1 using the CONT layer.

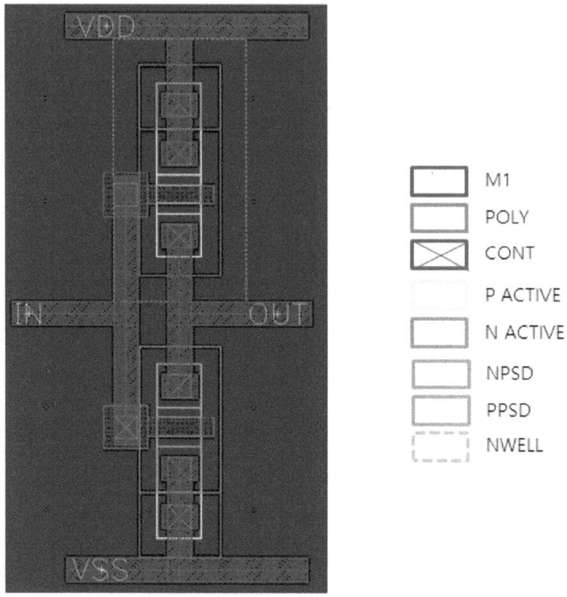


(a) in case that the top metal is M1

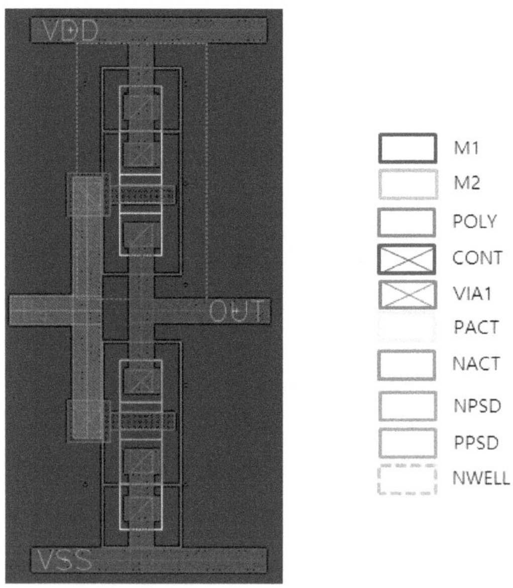


(b) in case that the top metal is M2

Fig. 3. Layout images of CMOS inverter.

The layout of CMOS inverter with top metal M2 is the same as Fig. 3(a) up to N-Well, P-Well formation, gate poly formation, S/D diffusion formation, contact hole formation, and M1 formation. In Fig. 3(a), the IN signal, M2, is connected to M1 by using the VIA1 layer, which is connected to the gate POLY of the NMOS and PMOS transistors using the CONT layer. The n+ diffusion, which is the drain of the NMOS transistor, and the p+ diffusion, which is the drain of the PMOS transistor, can output to the OUT signal through M1 using the CONT layer.

Table 1. Layers used in the layout of CMOS inverter.

| Used Layer | Layer Name | Description |
|---|---|---|
| | M1 | 1st Metal |
| | M2 | 2nd Metal |
| | POLY | Poly |
| | CONT | Contact |
| | VIA1 | Connects M1 and M2 |
| | PACT | P-Active |
| | NACT | N-Active |
| | NPSD | N+ Implantation |
| | PPSD | P+ Implantation |
| | NW | N-WELL |

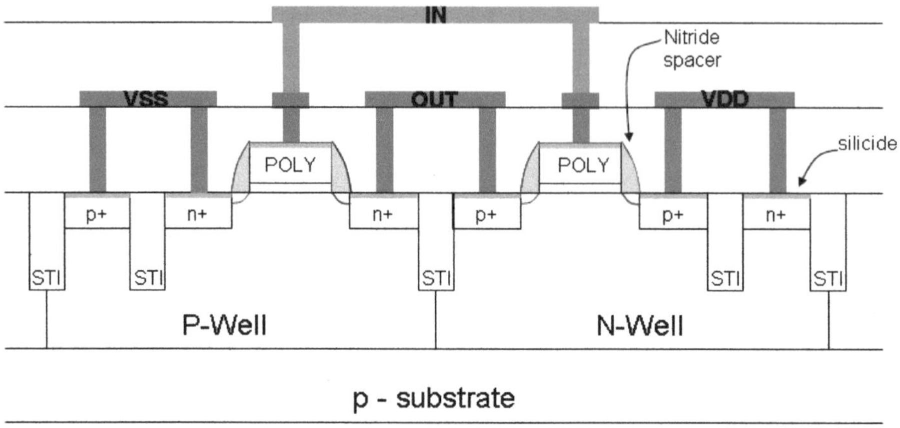


Fig. 4. Process cross-sectional view of CMOS inverter circuit with top metal M2.

A mask is fabricated from the semiconductor layout and a semiconductor IC is made through the fabrication process using the mask. The mask is a flat glass plate divided into transparent parts that pass light in a photo-lithography process and opaque parts that block light, and the patterns of transparent and opaque parts are determined according to the GDS data laid out by the designer. Table 1 shows the mask used in the twin-Well CMOS process and the layout layers used, AND where top metal is M2.

The process sequence for forming the twin-Well CMOS process cross-sectional view of Fig. 1 is shown in Fig. 5. First, column III acceptor impurities are injected to form the P-Well, and N-Well is implanted with the donor impurities of the column V. Next, shallow trench isolation (STI) of thick oxide is formed for electrical isolation between MOS devices. In CMOS process technology of 0.18$\mu$m or less, STI process technology is used instead of FOX (Field Oxide). The field area is the area other than the active area. After forming the gate oxide by thermal oxidation, a poly-silicon gate pattern is formed by using a poly mask, and source and drain diffusions of LDD (Lightly Doped Drain) structure are formed. In order to reduce the source, drain, and gate resistance, contact holes are formed using a CONT mask, and then M1 patterns are formed using an M1 mask after a silicide is formed and a CVD (Chemical Vapor Deposition) oxide is deposited,

Table 2. The mask used in the twin-Well CMOS process and the layout layers used.

| MASK ID | Layout Layer | Field | Remark |
|---|---|---|---|
| STI | NACT OR PACT | clear | Active Area |
| PW | NOT(NWELL) | dark | P-type well |
| NW | NWELL | dark | N-type well |
| POLY | POLY | clear | Gate Poly |
| N | NPSD | dark | N+S/D Implant |
| P | PPSD | dark | P+S/D Implant |
| CONT | CONT | dark | Contact |
| M1 | M1 | clear | Metal 1 |
| VIA1 | VIA1 | dark | Via 1 |
| M2 | M2 | clear | Metal 2 |

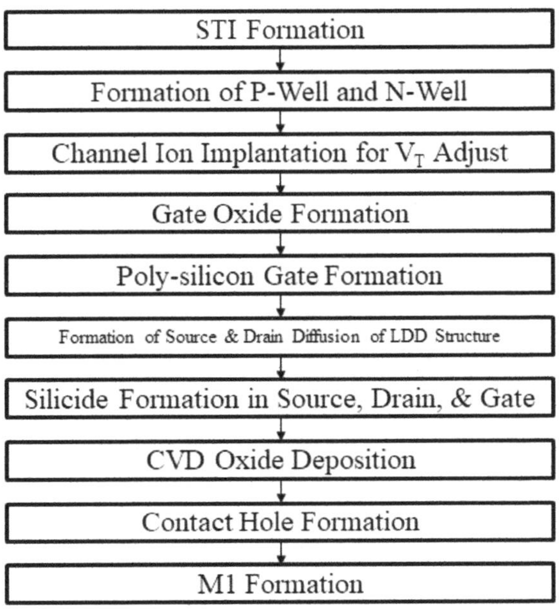


Fig. 5. Flow of twin-Well CMOS process.

## 4.2 Resistor Elements

The resistance R in the resistor of Fig. 6(a) is proportional to the length L and inversely proportional to the area A (= WT), where W is the width, T is the thickness, and the proportionality constant $\rho$ is the resistivity having a unit of $\Omega \cdot$cm. Eq. (1) can be written as Eq. (2). In Eq. (2), the sheet resistance Rsh is given by Eq. (3) and the unit is $\Omega$/sq. or $\Omega/\square$. Sheet resistance Rsh is the resistance between node C and node D when the length of the resistor is equal to the width, as shown in Fig. 6(b). In Eq. (2), the resistance is the number of sheets or squares, which is the length L divided by the width W. So the resistance R can be calculated by multiplying the sheet resistance by the number of sheets. For Rsh=50 $\Omega$/sq, W=2 ㎛, and L=1,000 ㎛, the resistance is 25 ㏀, as calculated in Eq. (4). On the other hand, for Rsh=50 $\Omega$/sq., the length L for realizing the resistance of 10 ㏀ may be 400 ㎛ with W=2 ㎛ as in Eq. (5).

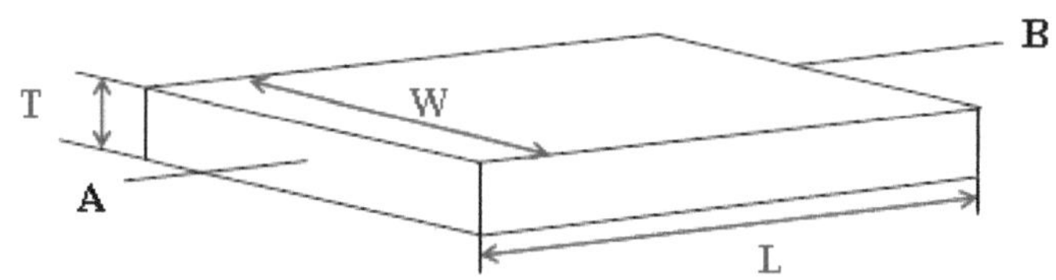


(a) geometrical structure

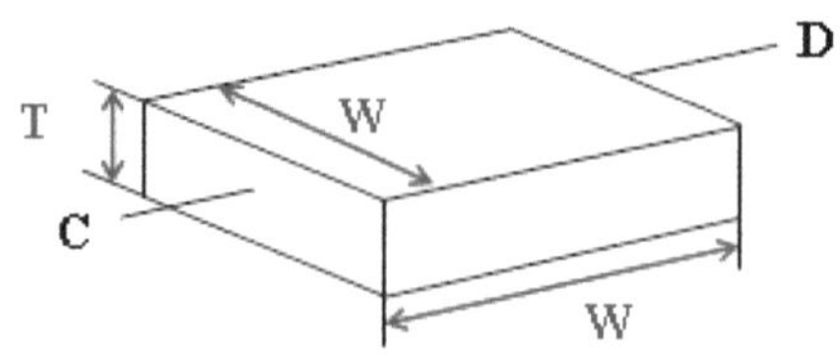


(b) sheet resistance

Fig. 6. Resistor.

$$R_{AB} = \rho \frac{L}{WT} \tag{1}$$

$$R_{AB} = \rho \frac{L}{WT} = \frac{\rho}{T} \cdot \frac{L}{W} = R_{sh} \cdot \frac{L}{W} \tag{2}$$

$$R_{sh} = \frac{\rho}{T} \tag{3}$$

$$R = R_{sh} \cdot \frac{L}{W} = 50\ \Omega/sq. \times \frac{1{,}000\ \mu m}{2\ \mu m} = 25\ K\Omega \tag{4}$$

$$L = \frac{W}{Rsh} \times R = \frac{2\ \mu m}{50\ \Omega/sq.} \times 10\ K\Omega = 400\ \mu m \tag{5}$$

Table 3. Comparison of characteristics of resistive elements that can be realized in CMOS integrated circuit process.

| element types | sheet resistances | characteristics |
|---|---|---|
| poly-silicon resistor | 100 Ω/□ ~ 10 kΩ/□ | • By lowering the doping concentration of poly-silicon, high-R resistance of several kΩ/□ can be achieved. |
| diffusion resistor | 50 Ω/□ ~ 150 Ω/□ | • big parasitic pn junction capacitance |
| N-Well resistor | typical 1 kΩ/□ | • big parasitic well junction capacitance<br>• big process variation |

The linear resistive elements used in CMOS analog integrated circuit design using CMOS integrated circuit process have a poly-silicon resistor, a diffusion resistor, and a well resistor as shown in Table 2. In the CMOS process, the gate and n+ and p+ diffusion of the MOS transistor have a sheet resistance of 15 Ω/sq.. The salicide metal is deposited by self-aligning process. On the other hand, the use of non-salicide resistors when using poly-silicon or diffusion as a resistor is more costly than the resistance of a salicide structure, and thus the non-salicide structure is advantageous for design.

The poly-silicon resistors such as NPOLY resistors, non-sal. PPOLY resistor and high-R ppoly resistor are shown in Fig. 7. High-R PPOLY resistor is doped with column III Boron in the same way as non-sal. PPOLY resistor but the doping concentration is lower by one order of PPOLY resistor with one additional mask and thus the sheet resistance is usually 1 kΩ/sq. ~ 10 kΩ/sq.. On the other hand, non-sal. PPOLY resistance is 300 Ω/sq. ~ 400 Ω/sq. which is lower than that of high-R PPOLY.

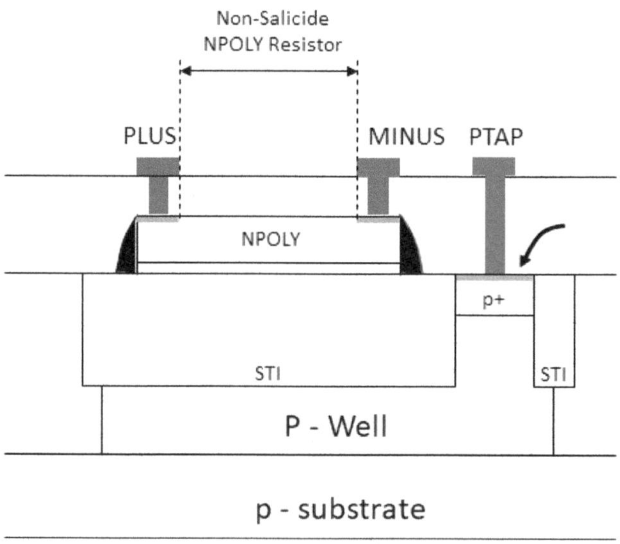


(a) non-sal. NPOLY resistor

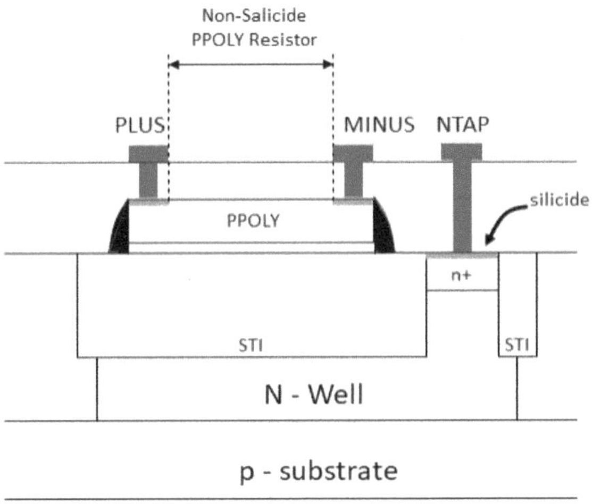


(b) non-sal. PPOLY resistor

Fig. 7. Structures of poly resistor.

Diffusion resistors such as non-sal. NDIFF resistor and non-sal. PDIFF resistor are shown in Fig. 8. The NDIFF resistor is a device in which n+ diffusion is formed in the P-Well, and the PDIFF resistor is a device in which p+ diffusion is formed in the N-Well. The DIFF resistive region are processed by the non-salicide and thus the sheet resistance is increased.

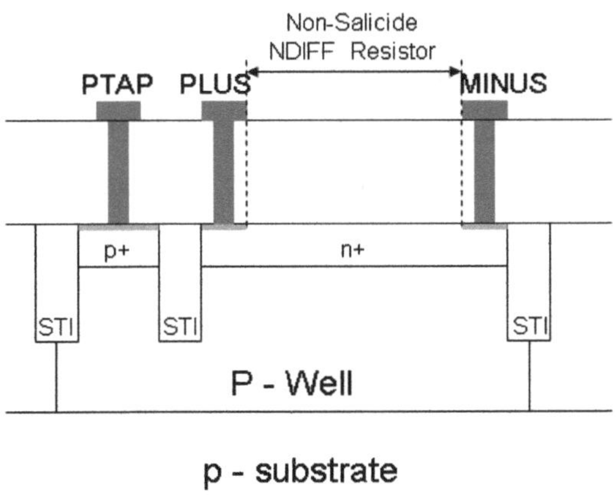


(a) non-sal. NDIFF resistor

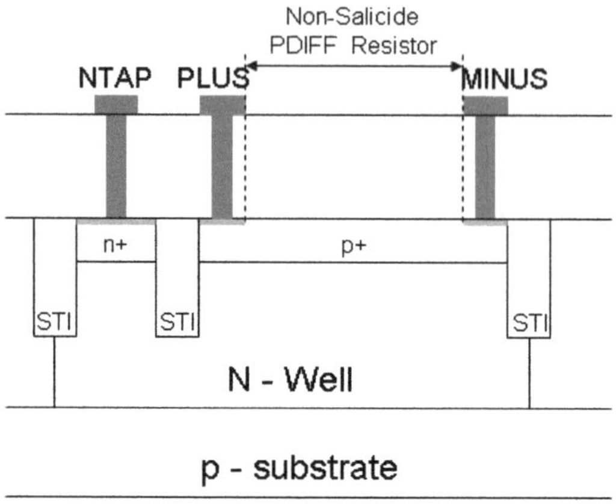


(b) non-sal. PDIFF resistor

Fig. 8. Structures of diffusion resistors.

## 4.3 Capacitor Elements

A plate capacitor is a case where there is an insulator of thickness t between two sheets of planar conductors with area A as shown in Fig. 9. The capacitance C between the two plates is given by Eq. (6), and the capacitance is proportional to the area A of the plate and inversely proportional to the thickness t. As shown in the plate capacitor, if there is a nonconductor between the two plate conductors, a capacitor is formed.

$$C = \frac{Q}{V} = \epsilon_0 \epsilon_r \frac{A}{t} \qquad (6)$$

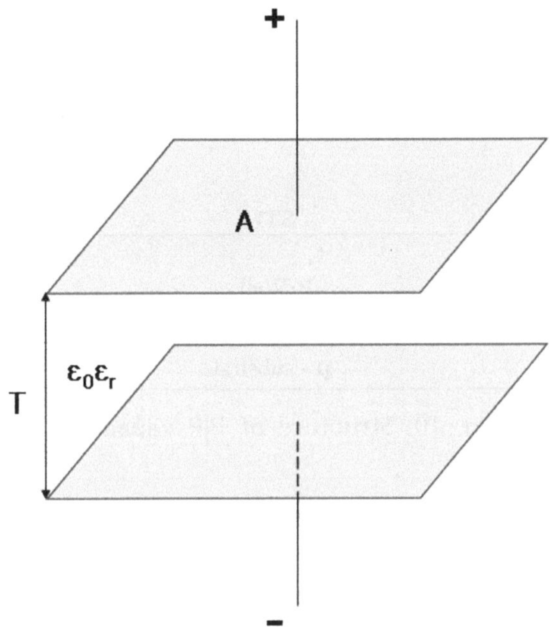


Fig. 9. Structure of plate capacitor.

In CMOS analog integrated circuit designs, capacitors are used in A/D converters, operational amplifiers, voltage regulators, and charge pump circuits. The capacitor has low voltage dependency and the larger the capacitance per unit area, the smaller the layout area, so the capacitor performance is excellent. The capacitor element may be a poly-insulator-poly (PIP) capacitor, a metal-insulator-metal (MIM) capacitor, a metal-oxide-metal (MOM) capacitor, or a MOS capacitor. PIP capacitors and MIM capacitors are close to ideal dielectric capacitors with almost no change in the capacitance depending on the applied voltage.

As shown in Fig. 10, the bottom plate of the PIP capacitor is POLY1 and the top plate is POLY2. Between POLY1 and POLY2 there is an insulating material of $SiO_2$. On the other hand, PIP capacitors require an additional POLY2 process.

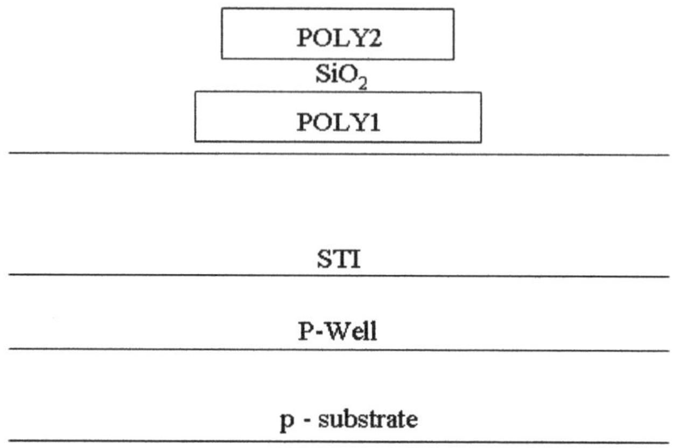


Fig. 10. Structure of PIP capacitor.

The MIM capacitor consists of M1 as the bottom plate and CTE(Capacitor Top Electrode) as the top plate as shown in Fig. 11(a). There is an insulating material of $SiO_2$ between M1 and CTE. The capacitance of the MIM capacitor is inversely proportional to the oxide thickness $t_{ox}$ between M1 and CTE, independent of the applied voltage, as shown in Eq. (7), and is proportional to the overlap area A of M1 and CTE shown in Fig. 11(b). The proportional constant is proportional to the product of the dielectric constant $\epsilon_0$ (=$8.854\times10^{-14}$ F/cm) in vacuum and $\epsilon_{ox}$ (=3.9). The MIM capacitance is provided by capacitors of 0.5 fF/$\mu m^2$, 1 fF/$\mu m^2$, 1.5 fF/$\mu m^2$, 2 fF/$\mu m^2$, and the like. On the other hand, the MIM capacitor requires additional CTE metal process.

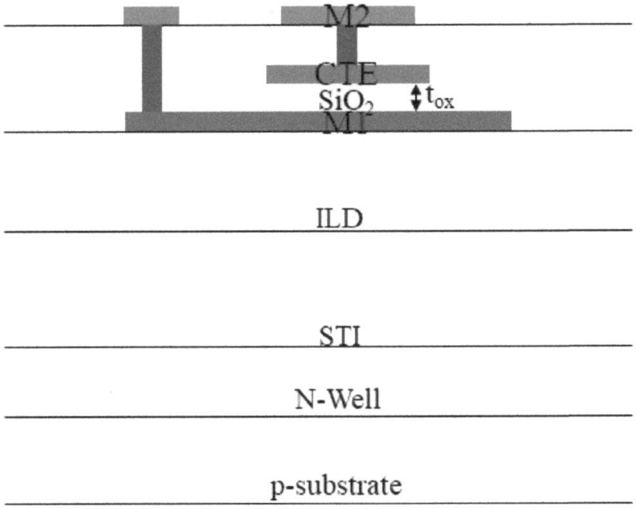


(a) cross-sectional view

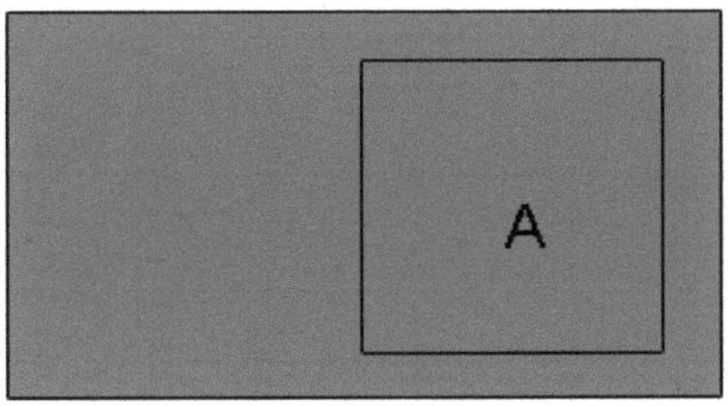


(b) floor plan

Fig. 11. Structure of MIM capacitor.

$$C = \epsilon_0 \epsilon_{ox} \frac{A}{t_{ox}} \tag{7}$$

MIM capacitors require additional CTE metal process, while metal-on-metal (MOM) capacitors can be implemented using a metal layer provided in a standard CMOS process, as shown in Fig. 12, using a woven structure, an interdigitated parallel wire structure, a parallel stacked wire structure, and a stack MIM capacitor.

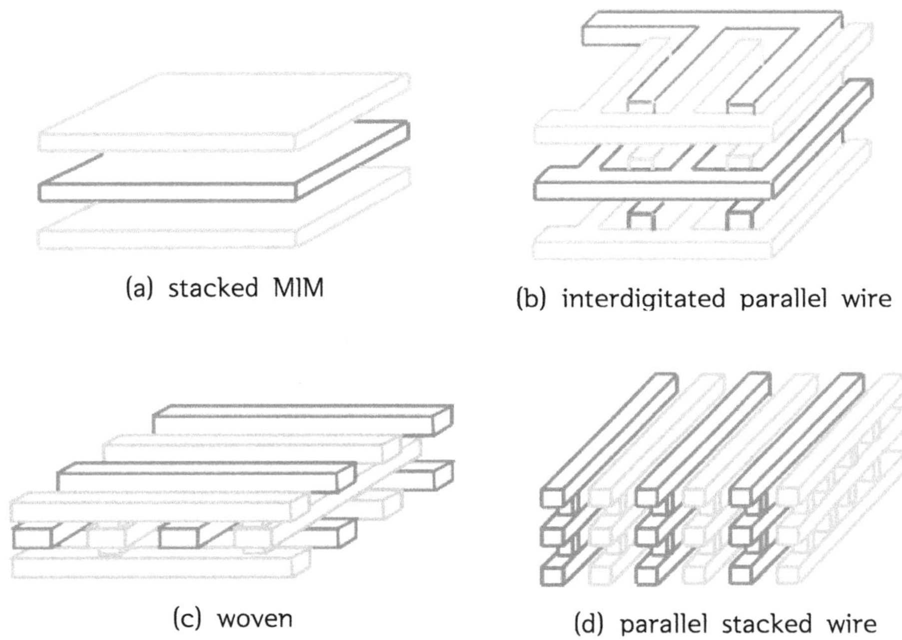


Fig. 12. Structures of MOM capacitor.

# Exercises

1. Answer the following questions.

   (a) What does CMOS stand for?

   (b) Draw an inverter circuit using an open drain.

2. Answer the following questions for the CMOS inverter.

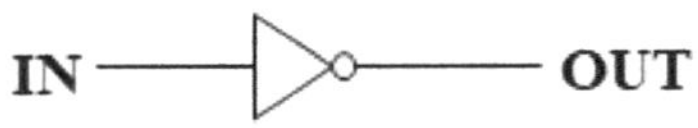


   (a) Draw the circuit diagram of transistor level. Mark VDD, VSS, and the body of MOS transistor.

   (b) Draw the process cross-sectional view.

3. Draw the process cross-sectional view of the twin-well CMOS.

4. Draw the process cross section when the bias voltage of the NMOS transistor is applied as shown below. Here, $V_{TN}$= 0.5 V.

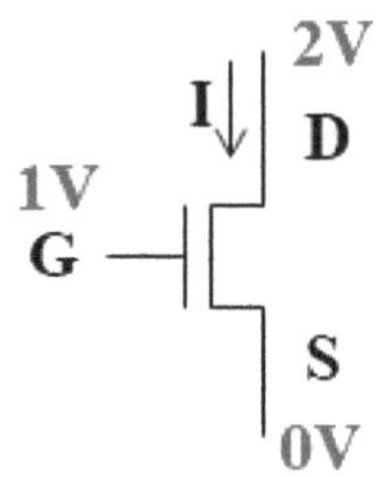


5. Draw the process cross section when the bias voltage of the PMOS transistor is applied as shown below. Here, $V_{TP}$= −0.5 V.

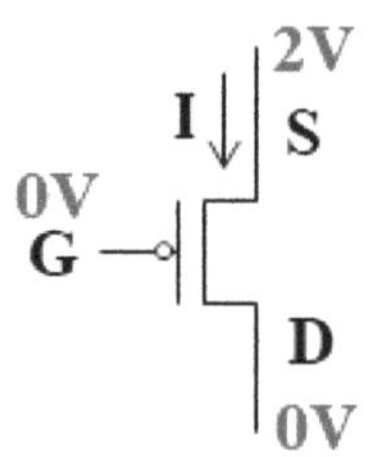


6. If $\mu_p C_{ox}$ = 35 μA/V$^2$, W=2 μm, $V_{GS}$ = −2.5 V, and $V_{TP}$ = −0.5 V, calculate the length L to have a PMOS linear resistance of $R_P$=10 kΩ.

7. If $R_{sh}$=1 kΩ/sq and W=2 μm, what is the length L of polysilicon to realize 100 kΩ resistance?

8. When using a MIM capacitor of 1 fF/μm$^2$, calculate the area to have a capacitance of 1pF.

9. Answer the following questions. Here, W=0.5 μm, L=1,000 μm, T=5,000 Å, H=10,000 Å, and $R_{sh}$ = 0.1 Ω/sq.

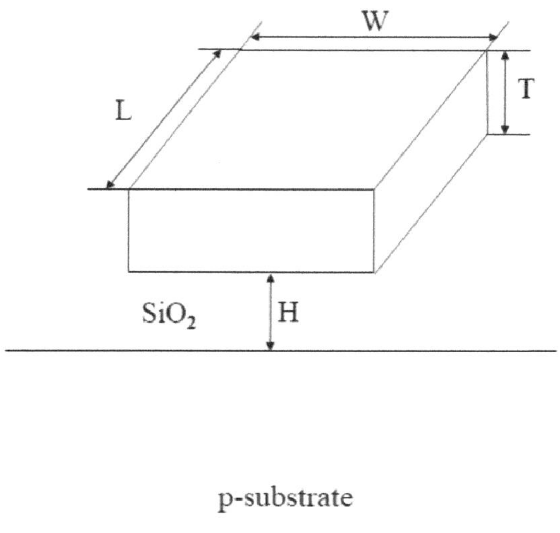


(a) Calculate the R of the interconnect line.

(b) Calculate the intrinsic capacitance $C_I$ of the interconnect line using the following formula.

$$C_I = \epsilon_0 \epsilon_{ox} \left\{ 1.15 \times \frac{W}{H} + 2.8 \times \left(\frac{T}{H}\right)^{0.222} \right\} \times L$$

10. Answer the following questions. Here, W=0.5 μm, L=10,000 μm, T=5,000 Å, H=10,000 Å, S=0.5 μm, and $R_{sh}$ = 0.1 Ω/□.

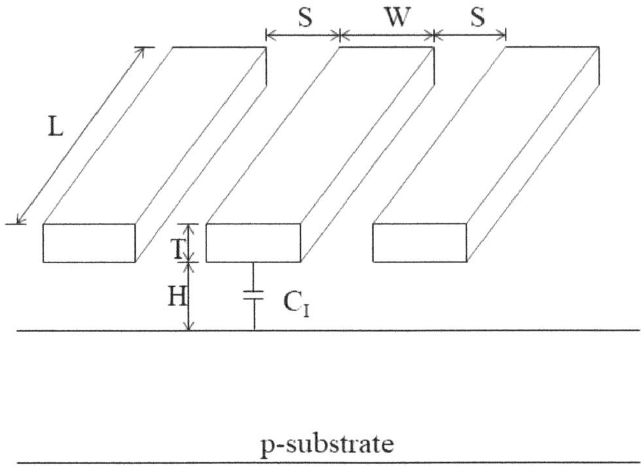


(a) Calculate the R of the interconnect line.

(b) Calculate the intrinsic capacitance $C_I$ of the interconnect line using the following formula.

$$C_I = \epsilon_0 \epsilon_{ox} \left\{ 1.15 \times \frac{W}{H} + 2.8 \times \left(\frac{T}{H}\right)^{0.222} \right\} \times L$$

(c) Calculate coupling capacitance $C_c$ of the interconnect line using the following formula.

$$C_C = \epsilon_0 \epsilon_{ox} \left\{ 0.03 \times \frac{W}{H} + 0.83 \times \frac{T}{H} - 0.07 \times \left(\frac{T}{H}\right)^{0.222} \right\} \times \left(\frac{S}{H}\right)^{-1.34} \times L$$

(d) Calculate the total capacitance $C_T$ of the middle interconnect line.

# Chap. 4 References


[1] Kyungwook Shin, Design of CMOS Digital Integrated Circuits, Hanbit Academy, 2014.

[2] https://ko.wikipedia.org/wiki/%EC%A7%91%EC%A0%81_%ED%9A%8C%EB%A1%9C_%EC%84%A4%EA%B3%84.

[3] Hongjun Park, Design of CMOS Analog Integrated Circuits, $\Sigma$ Sigma Press, 1999.

[4] Byungi Son et al., Electromagnetics, Hakmunsa, 1985.

[5] Kiseong Son, Design of Analog Semiconductor Circuits, Bokdu Press, 2017.

[6] N. C. Chen et al., "High-Density MOM Capacitor Array with Novel Mortise-Tenon Structure for Low-Power SAR ADC," Proceedings of 2017 Design, Automation and Test in Europe, pp. 1757-1762, March 2017.

[7] T. Sakurai et al., " Simple Formulas for Two- and Three Dimensional Capacitances, " IEEE Transactions on Electron Devices, vol. ED-30, no. 2, pp. 183-185, Feb. 1983.

# 5 CHAPTER Single-Stage Amplifiers

## 5.1 Small-Signal Equivalent Circuits

Amplification is essential in most analog circuits. A single-stage MOS amplifier circuit is a basic circuit of a multi-stage amplifier and an OPAMP circuit, and has a common source amplifier, a common gate amplifier, and a common drain amplifier.

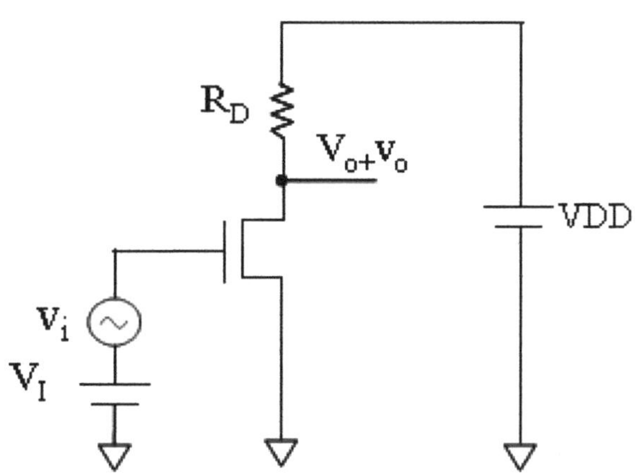


(a) circuit

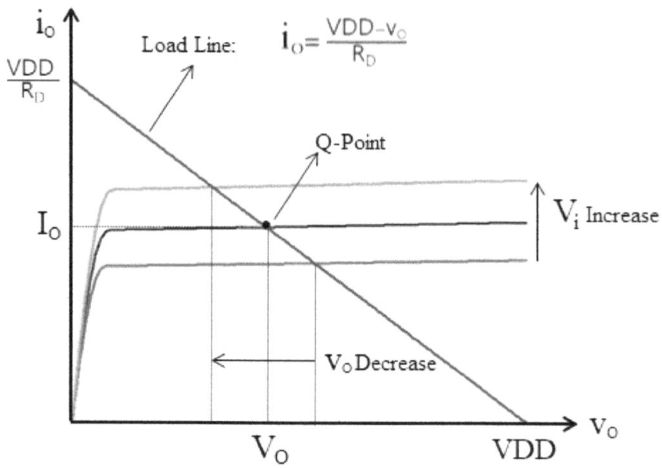


(b) operating point (quiescent point)

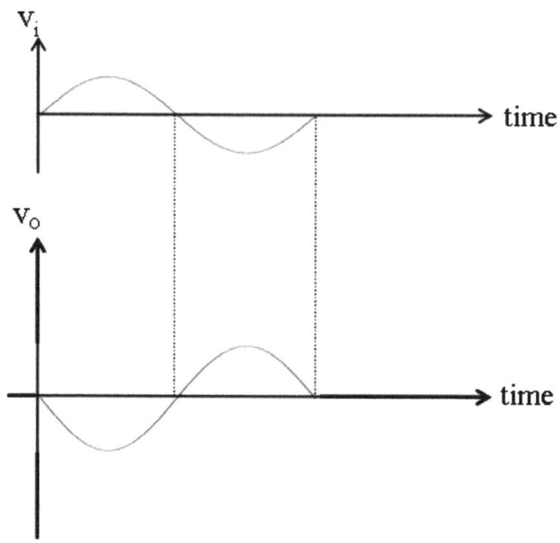


(c) small-signal voltage waveform

Fig. 1. Common source amplifier.

In the common source amplifier circuit of Fig. 1, the source node is connected to the ac ground, the input signal is connected to the gate node, and the output signal is output from the drain node. The ac ground means that the signal of the corresponding node is time invariant. A resistor $R_D$ is connected between the drain of the MOS transistor and the power supply voltage of VDD. The MOS transistor converts the change of the gate-source voltage $v_i$ to the small signal drain current $i_{ds}$ by the transconductance, and the small signal output voltage $v_o$ occurs when the $i_{ds}$ current flows through the $R_D$.

The large-signal input voltage $v_I$ and the output voltage $v_O$ are shown in Eqs. (1) and (2).

$$v_I = V_I + v_i \tag{1}$$

$$v_O = V_O + v_o \tag{2}$$

where $V_I$ and $V_O$ represent the DC input voltage and the DC output voltage at the DC operating point, and $v_i$ and $v_o$ represent the small-signal input voltage and the small-signal output voltage, respectively.

Fig. 2 shows $V_I$=1.2 V, VDD=5 V, and $R_D$=20 kΩ in a common source amplifier circuit using a 0.13 μm CMOS process. The size of the NMOS transistor is W = 10 μm, L = 2 μm, and M = 3. M is a multiplier, meaning that there are three unit transistors of 10 μm/2 μm, connecting gate to gate, drain to drain, source to source, and body to body. If large-signal input voltage $v_I$ is applied at 1.19 V, 1.2 V, and 1.21 V, $i_O$ and $v_O$ are outputs as shown in Table 1. When $v_I$=1.2 V, then $i_O$ and $v_O$ are 148 μA and 2.05 V, respectively.

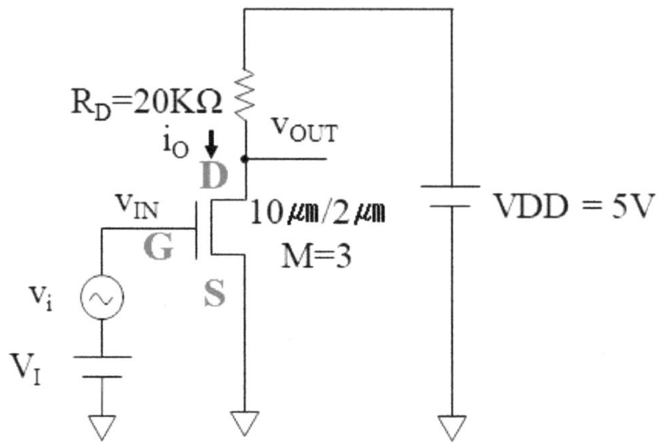


(a) circuit

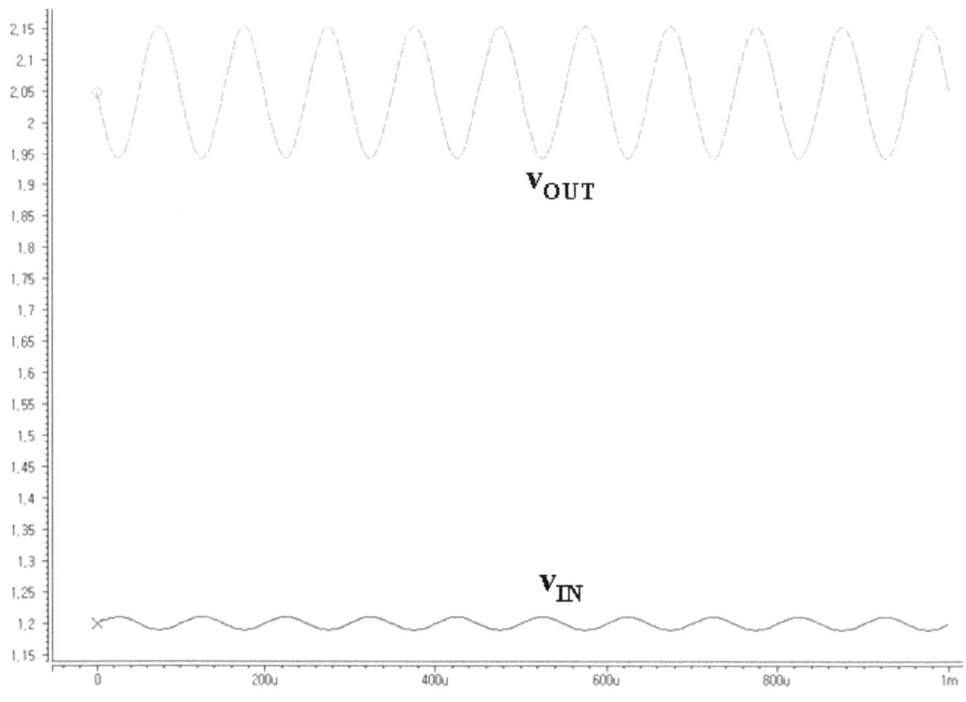


(b)

Fig. 2. A common source amplifier using 0.13 μm CMOS process.

Table 1. Output current $i_O$ and output voltage $v_O$ according to input voltage $v_I$.

| $v_I(=v_{GS})$ | $i_O(=i_{DS})$ | $v_O(=v_{DS})$ |
|---|---|---|
| 1.19 V | 142 μA | 2.16 V |
| 1.2 V | 148 μA | 2.04 V |
| 1.21 V | 153 μA | 1.94 V |

The DC operating point is the point where $v_I = V_I$ and $v_O = V_O$ when $v_i = 0$ and $v_o = 0$, and is the point where the $i_{DS} - v_{DS}$ characteristic curve of the NMOS transistor used in the common source amplifier circuit as shown in Fig. 1(b) meets the load line of Eq. (3).

Eq. (3) is derived by applying the KVL at the output loop as shown in Fig. 2(a). The $i_O$ curve with respect to $v_O$ when $v_I = V_I$ shown in Fig. 1(b) is the $i_{DS} - v_{DS}$ characteristic curve of the NMOS transistor shown in Fig. 2(a). And the load line shown in Fig. 1(b) has a slope of -1/$R_D$ and an absolute current of $VDD/R_D$.

$$i_O = -\frac{v_O}{R_D} + \frac{VDD}{R_D} \tag{3}$$

As shown in Fig. 1(c), when a sinusoidal wave is applied with the small-signal input voltage $v_i$ at the large-signal input voltage $v_I$, the small-signal output voltage $v_o$ has the opposite sign and appears as the small-signal output voltage amplified by the voltage gain.

In the amplifier circuit on the Fig. 1(a), assuming that the large-signal input voltage $v_I$ is applied to the gate of the MOS transistor and the MOS transistor operates in the saturation region, the large-signal output current $i_O$ is expressed by Eqs. (4) and (5). When the Eqs. (4) and (5) are solved together, the small-signal output voltage $v_o$ is given by Eq. (6).

$$i_O = \frac{1}{2}\mu_n Cox\frac{W}{L}(V_I + v_i - V_{TN})^2 \tag{4}$$

$$i_O = \frac{VDD - (V_O + v_o)}{R_D} \tag{5}$$

$$v_o = -R_D \cdot \mu_n Cox \cdot \frac{W}{L}(V_I - V_{TN})v_i\left\{1 + \frac{v_i}{2(V_I - V_{TN})}\right\} \tag{6}$$

As shown in Eq. (6), when the small-signal input voltage $v_i$ is large, the small-signal output voltage $v_o$ exhibits a nonlinear characteristic with respect to $v_i$ and a signal distortion occurs. In order to make the signal distortion rate within 10%, $v_i$ voltage must satisfy Eq. (7).

$$|v_i| \le 0.2(V_I - V_{TN}) \tag{7}$$

If we neglect the term $v_i^2$ in Eq. (6), the small-signal output voltage $v_o$ is linearly proportional to the small-signal input voltage $v_i$ as shown in Eq. (8).

$$v_o \cong -R_D \mu_n Cox\frac{W}{L}(V_I - V_{TN})v_i \tag{8}$$

For small-signal circuit analysis of an MOS amplifier circuit, a small-signal equivalent model of an MOS transistor element used in an amplifier circuit is required. Fig. 3 shows a hybrid-$\pi$ equivalent model with a low-frequency small-signal equivalent model of an NMOS transistor. The NMOS transistor in Fig. 3(a) acts as a voltage controlled current source that receives the voltage $v_{gs}$ of the gate and source and supplies a dependent current source $g_m v_{gs}$ between the drain and the source. The input resistance between the gate and the source is infinite, and there is a small-signal output resistance $r_o$ between the drain and the source.

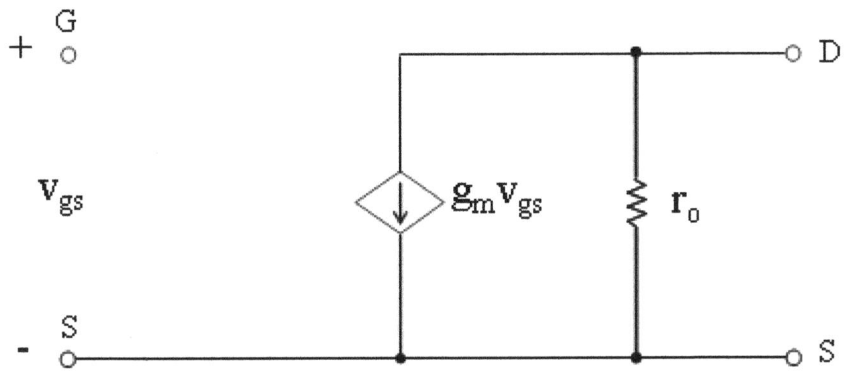


(a) NMOS transistor

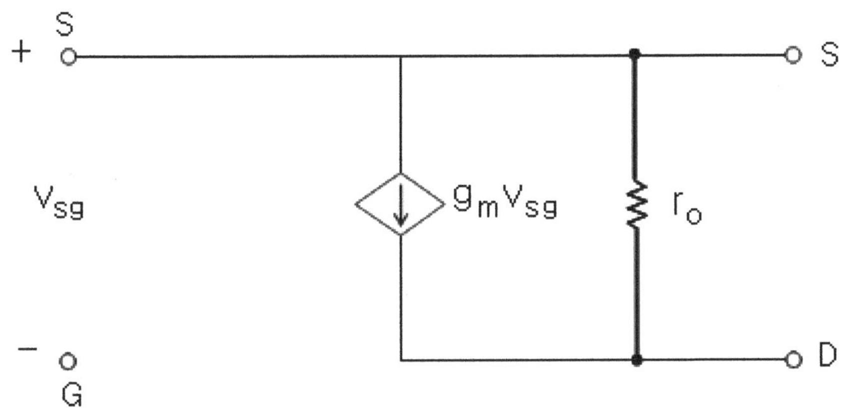


(b) PMOS transistor

Fig. 3. Small-signal equivalent circuit of a MOS transistor.

When the NMOS transistor is in the saturation region, the drain current increases linearly with the channel length modulation effect as the $v_{DS}$ voltage increases as shown in Eq. (9).

$$i_D = \frac{1}{2}\mu_n Cox\frac{W}{L}(v_{GS} - V_{TN})^2 \cdot (1 + \lambda v_{DS}) \tag{9}$$

The small signal model parameter $g_m$ is obtained from Eq. (9) where $\lambda$ is set to zero. And $r_o$ is obtained from Eq. (9). The transconductance $g_m$ at the DC operating point is given by Eq. (10). As shown in Eq. (10), $g_m$ is proportional to the square root of the DC drain current $I_D$ and is proportional to the square root of W/L for a given drain current $I_D$.

$$g_m \equiv \frac{\partial i_D}{\partial v_{GS}}\Big|_{v_{GS} = V_{GS}} = \sqrt{2\mu_n C_{OX}\frac{W}{L}I_D} = \frac{I_D}{\frac{V_{GS} - V_{TN}}{2}} \tag{10}$$

The small-signal output resistance $r_o$ is shown in Eq. (11). $r_o$ equals inversely proportional to the equation in the bracket. Eq. (11) is around $i_D$ over around $v_{DS}$ when the instantaneous voltage $v_{GS}$ equals to the DC voltage $v_{GS}$.

$$r_o \equiv \left\{ \frac{\partial i_D}{\partial v_{DS}}\Big|_{v_{GS} = V_{GS}} \right\}^{-1} = \frac{1}{\lambda I_D} \tag{11}$$

## 5.2 Characteristics of Amplifiers

Fig. 4 shows a typical amplifier circuit consisting of signal voltage $v_s$, internal resistance $R_s$, and load resistance $R_L$. The signal source has a signal voltage $v_s$ and an internal resistance $R_s$, and the output node of the amplifier has a load resistance $R_L$. The load resistance $R_L$ is the input resistance of the next stage amplifier.

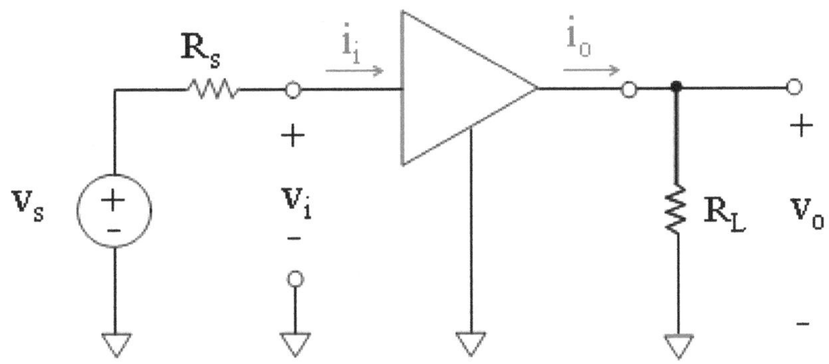


Fig. 4. A typical amplifier circuit consisting of signal voltage $v_s$, internal resistance $R_s$, and load resistance $R_L$.

Fig. 5 shows a typical amplifier circuit showing the amplifier as an equivalent circuit. The input resistance $R_i$ is obtained by the load effect of the amplifier on the signal source as shown in Eq. (12).

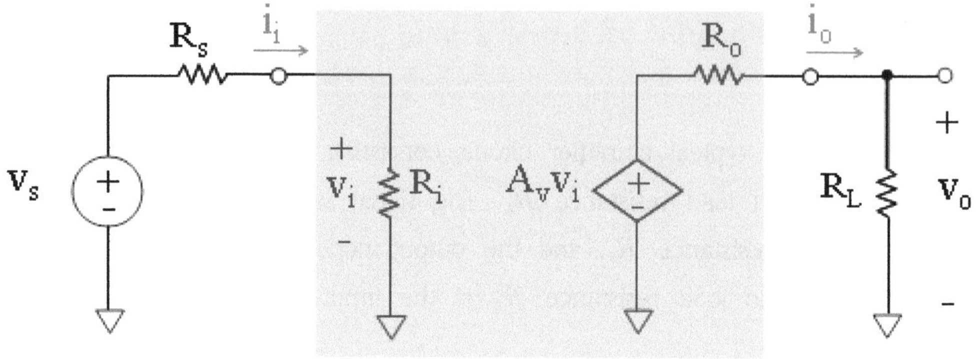


Fig. 5. A typical amplifier circuit in which the amplifier is represented by an equivalent circuit.

$$R_i \equiv \frac{v_i}{i_i} \tag{12}$$

The input voltage $v_i$ of the amplifier is determined by the resistance division of $R_s$ and $R_i$ of $v_s$ voltage as shown in Eq. (13).

$$v_i \equiv \frac{R_i}{R_s + R_i} v_s \tag{13}$$

$A_v$ in Fig. 5 represents the ratio of the output voltage $v_o$ to the input voltage $v_i$ when $R_L$=∞ as shown in Eq. (14) with open-circuit voltage gain.

$$A_v \equiv \frac{v_o}{v_i}|_{R_L = \infty} \tag{14}$$

The output resistance, $R_o$, is shown in Fig. 6, where $v_i$=0, and is calculated as shown in Eq. (15).

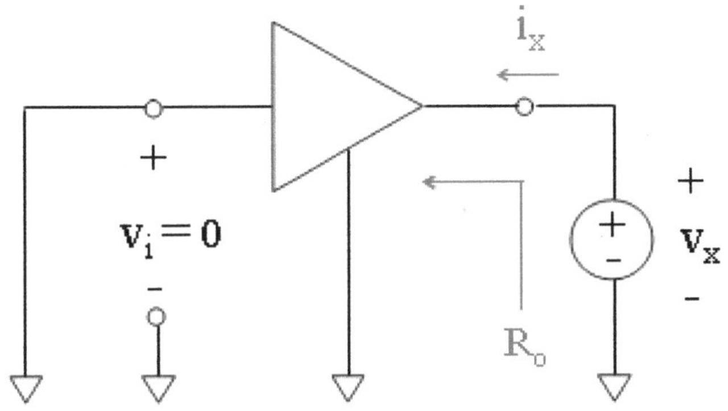


Fig. 6. A circuit for calculating the output resistance $R_o$ of a typical amplifier circuit.

$$R_o \equiv \frac{v_x}{i_x}\Big|_{v_i = 0} \tag{15}$$

The procedure for analyzing the MOS amplifier circuit is as follows.

① Calculating the $I_D$ of a MOS transistor in a DC equivalent circuit.

② Calculation of small-signal model parameters $g_m$ and $r_o$.

③ The DC voltage source is short, and the DC current source is open.

④ A small-signal equivalent circuit in which a MOS transistor is replaced with a small-signal equivalent model such as small-signal hybrid-$\pi$ equivalent model.

⑤ Small-signal equivalent circuit analysis such as voltage gain, input resistance, and output resistance, and so on.

## 5.3 Common Source Amplifier

In the common source amplifier circuit, the source node is connected to the ac ground, the input signal is connected to the gate node, and the output signal is output from the drain node. A resistor $R_D$ is connected between the drain of the MOS transistor and the voltage source VDD. In the common source amplifier circuit of Fig. 2(a), $I_D$=148 μA and $R_D$=20 kΩ, and as NMOS transistor related parameters $\mu_n C_{ox}$=70 μA/V$^2$, $V_A$=50 V, W=10 μm, L=2 μm, and M=3. Fig. 7 shows the DC equivalent circuit of a common source amplifier circuit for calculating DC operating points. As shown in Table 1, when $V_I$=1.2 V, the drain current $I_O$ flows 148 μA. Since $I_O$ is 148 μA, $g_m$ and $r_o$ are calculated as shown in Eqs. (16) and (17).

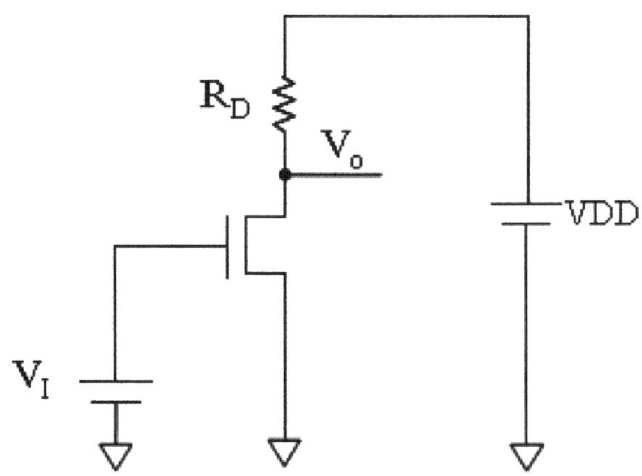


Fig. 7. DC equivalent circuit of a common source amplifier.

$$g_m = \sqrt{2\mu_n C_{OX}\frac{W}{L}I_D} = \sqrt{2\times 70\mu A/V^2 \times \frac{10\ \mu m \times 3}{2\ \mu m} \times 148\ \mu A} = 0.56\ mS \quad (16)$$

$$r_o = \frac{1}{\lambda I_D} = \frac{1}{0.02\ V^{-1} \times 148\ \mu A} = 338\ K\Omega \quad (17)$$

The ac equivalent circuit of the common source amplifier circuit can be obtained by shorting the voltage sources $V_I$ and VDD in the circuit of Fig. 2(a). The ac equivalent circuit of the common source amplifier circuit shown in Fig. 2(a) is as shown in Figs. 8 and 9 shows an ac equivalent circuit using a small-signal equivalent model, which is a hybrid-$\pi$ model of an NMOS transistor in the ac equivalent circuit of Fig. 8.

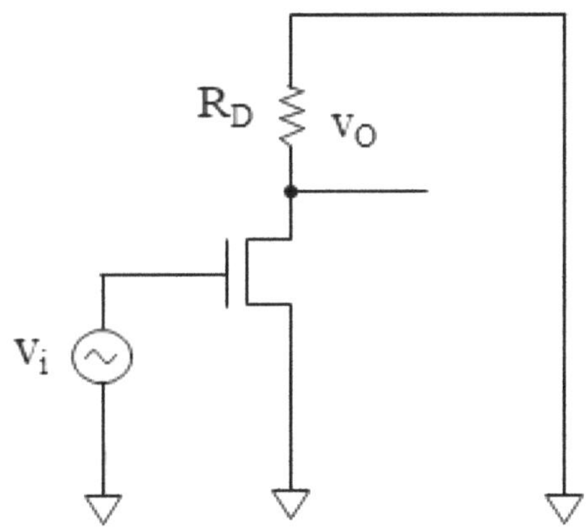


Fig. 8. ac equivalent circuit of a common source amplifier.

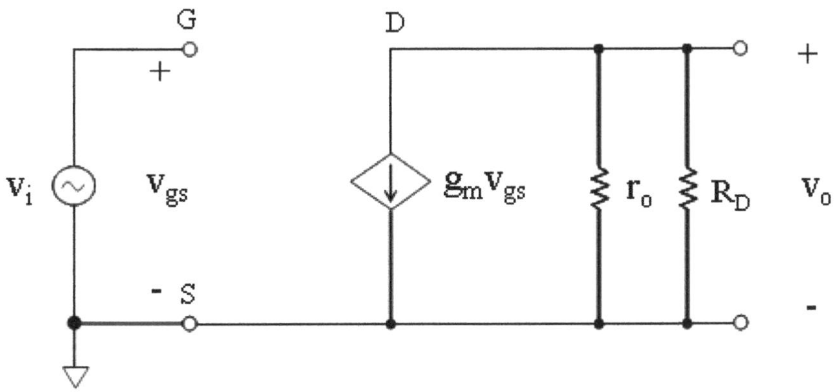


Fig. 9. ac equivalent circuit of a common source amplifier circuit applying a small-signal equivalent model.

When the KVL (Kirchhoff's Voltage Law) is applied to the output port, the small-signal output voltage $v_o$ is given by Eq. (18).

$$v_o = -g_m(r_o \| R_D)v_{gs} = -g_m(r_o \| R_D)v_i \tag{18}$$

And the low-frequency voltage gain $A_v$ is given by Eq. (19).

$$A_v = \frac{v_o}{v_i} = -g_m(r_o \| R_D) \tag{19}$$

Substituting $g_m$ in Eq. (16) and $r_o$ in Eq. (17) into Eq. (19), the voltage gain $A_v$ is -10.58. The small-signal input signal $v_i$ of the common source amplifier must satisfy the relation of Eq. (20) in order to make distortion less than 10%.

$$|v_i| \leq 0.2(V_{GS} - V_{TN}) \tag{20}$$

Fig. 10(a) shows an ac equivalent circuit for finding the small-signal input resistance $R_i$. It is a resistor that is seen from the input port when $R_L$ is infinite. As shown in Fig. 10(b), $R_i$ is infinite as shown in Eq. (21) because the input current $i_i$ is zero.

$$R_i = \frac{v_i}{i_i}\Big|_{R_L=\infty} = \infty \tag{21}$$

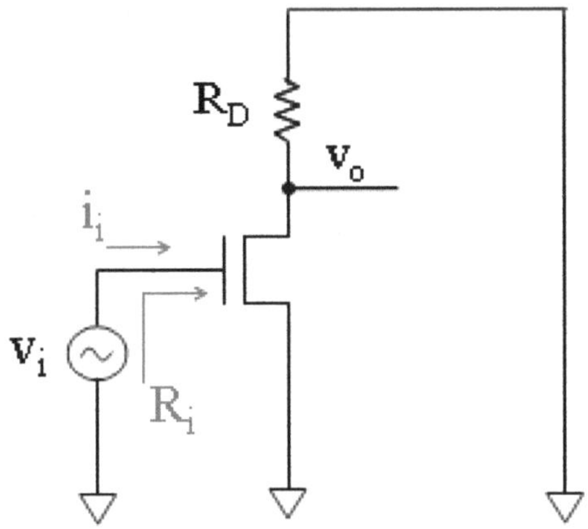


(a) ac equivalent circuit

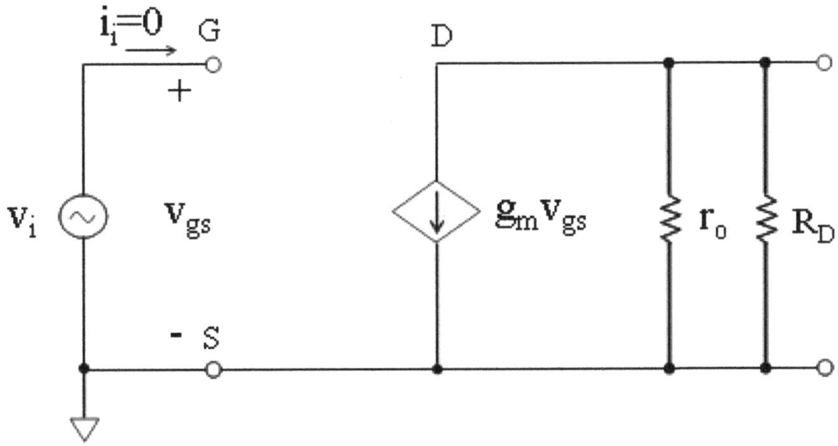


(b) ac equivalent circuit with small-signal equivalent model applied

Fig. 10. Circuit for finding small-signal input resistance $R_i$.

Fig. 11(a) is an ac-equivalent circuit for finding the small signal output resistance $R_o$. $R_o$ is obtained by finding the current $i_x$ flowing when $v_x$ is applied to the output port when $v_i$=0. As shown in the circuit of Fig. 11(b), since $v_i$=0, then $v_{gs}$ becomes 0 and the dependent current source $g_m v_{gs}$ between the drain and the source becomes zero.

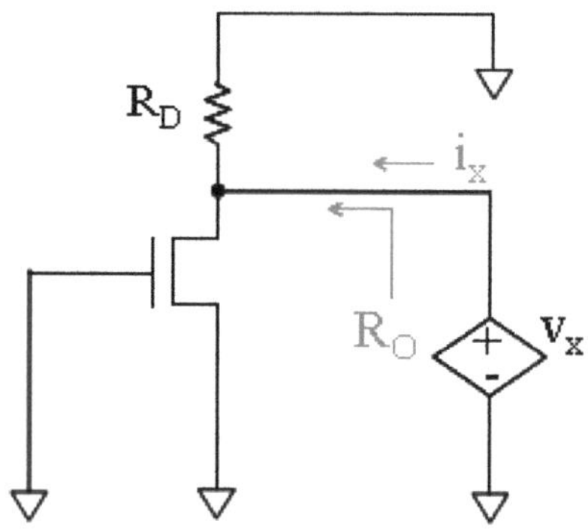


(a) ac equivalent circuit

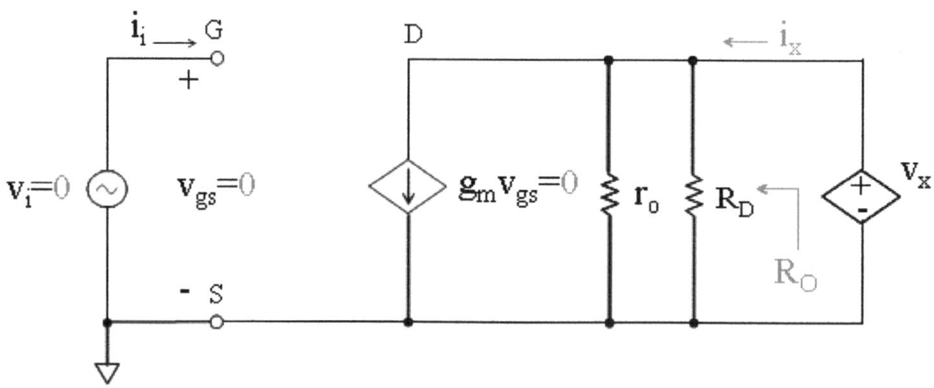


(b) ac equivalent circuit with small-signal equivalent model applied

Fig. 11. Circuit for obtaining small-signal output resistance $R_o$.

Fig. 12(a) shows the equivalent circuit when $g_m v_{gs}$=0, and $r_o$ and $R_D$ are parallel resistors, so it can be drawn as a simplified equivalent circuit as shown in Fig. 12(b). Equation (22) can be obtained by applying the KVL to the output port, and the small signal output resistance $R_o$ is calculated by $r_o \parallel R_D$ as shown in Eq. (23).

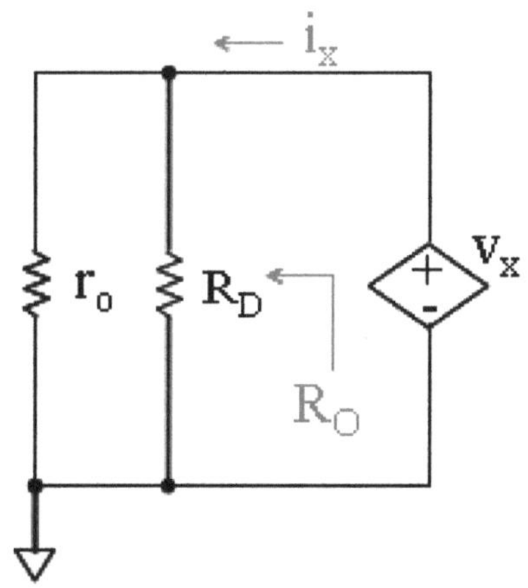


(a) equivalent circuit in case of $v_i$=0

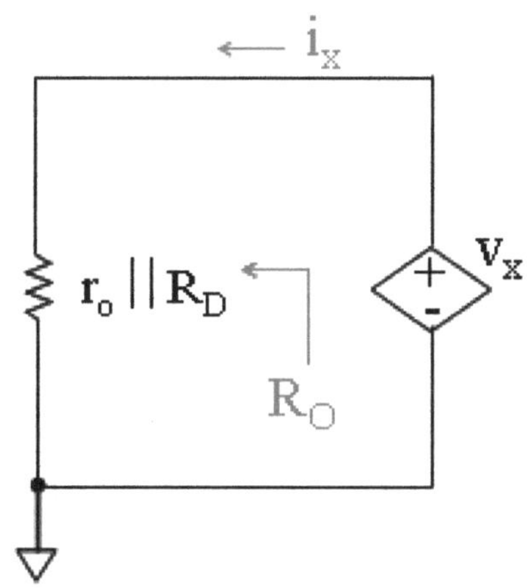


(b) simplified equivalent circuit with $r_o \parallel R_D$ applied.

Fig. 12. Simplified equivalent circuit for obtaining $R_o$.

$$v_x = (r_o \| R_D) \times i_x \tag{22}$$

$$R_o = \frac{v_x}{i_x}|_{v_i = 0} = r_o \| R_D = 18.9\ K\Omega \tag{23}$$

## 5.4 Common Drain Amplifier

In the common drain amplifier circuit of Fig. 13, the drain node is connected to the ac ground, the input signal is connected to the gate node, and the output signal is output from the source node. A resistor $R_S$ is connected between the source node of the MOS transistor and GND.

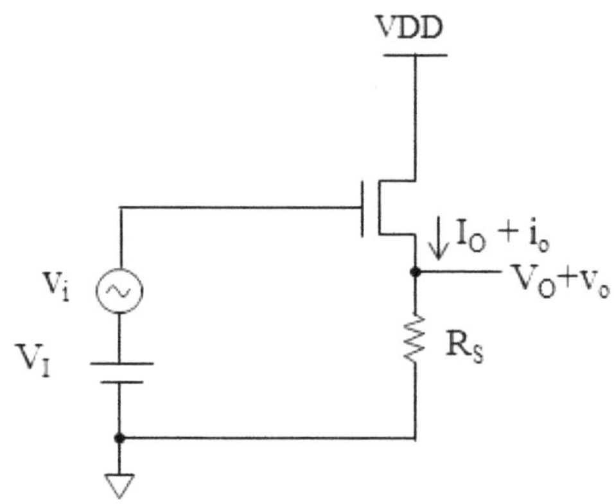


Fig. 13. Common drain amplifier.

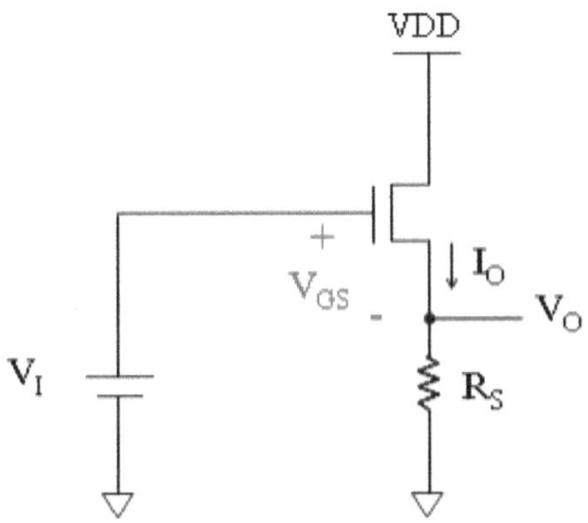


Fig. 14. DC equivalent circuit of a common drain amplifier.

Fig. 14 shows the DC equivalent circuit of a common drain amplifier circuit for calculating the DC operating point. We can get the DC equivalent circuit of the common drain circuit by putting $v_i$=0 and $v_o$=0.

The ac equivalent circuit of the common drain amplifier circuit shown in Fig. 15 can be obtained by shorting the DC voltage sources $V_I$ and VDD in the circuit of Fig. 13. Fig. 16 shows an ac equivalent circuit using a small-signal equivalent model, which is a hybrid-$\pi$ model of NMOS transistor in the ac equivalent circuit of Fig. 15. Applying the KVL to the input loop shown in Fig. 16 yields Eq. (24).

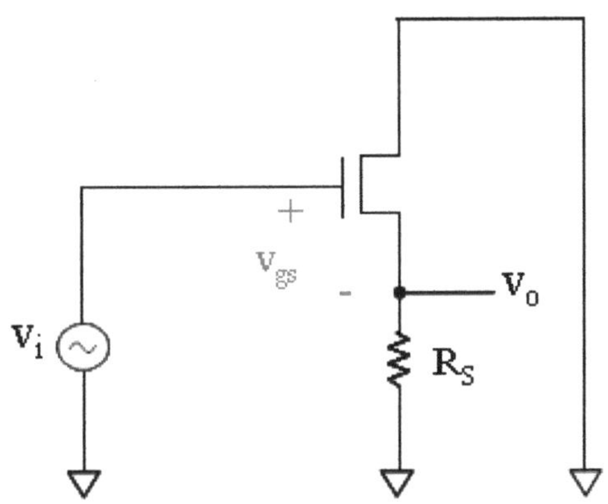


Fig. 15. ac equivalent circuit of common drain amplifier.

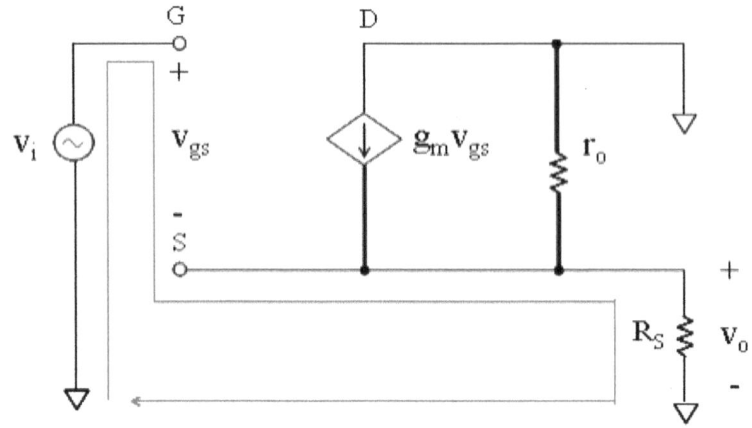


Fig. 16. ac equivalent circuit of common drain amplifier circuit with small-signal equivalent model applied.

$$-v_i + v_{gs} + v_o = 0 \tag{24}$$

Then, the KCL (Kirchhoff's Current Law) is applied to the source node of Fig. 16, we can obtain Eq. (25).

$$-g_m v_{gs} + \frac{v_o}{R_S} + \frac{v_o}{r_o} = 0 \tag{25}$$

In Eq. (25), $r_o$ and $R_S$ are parallel resistors and the resistance can be rewritten as in Eq. (26).

$$-g_m v_{gs} + \frac{v_o}{r_o \| R_s} = 0 \tag{26}$$

Eq. (26) is rewritten as Eq. (27).

$$v_{gs} = \frac{v_o}{g_m (r_o \| R_s)} \tag{27}$$

Substituting Eq. (27) into Eq. (24) yields Eq. (28).

$$-v_i + \frac{v_o}{g_m(r_o \| R_S)} + v_o = 0 \tag{28}$$

If we rewrite Eq. (28) for vi, it is as in Eq. (29).

$$v_i = \left(1 + \frac{1}{g_m(r_o \| R_S)}\right) \times v_o \tag{29}$$

Therefore, the voltage gain $A_v$ is given by Eq. (30).

$$A_v = \frac{v_o}{v_i} = \frac{1}{1 + \frac{1}{g_m(r_o \| R_S)}} \cong 1 \tag{30}$$

Since the small-signal parameter $g_m(r_o \| R_S) \gg 1$, the voltage gain $A_v$ of the common drain amplifier is almost 1, so it is used as a voltage buffer.

As shown in Fig. 17, since $i_i$=0, the small-signal input resistance $R_i$ is infinite.

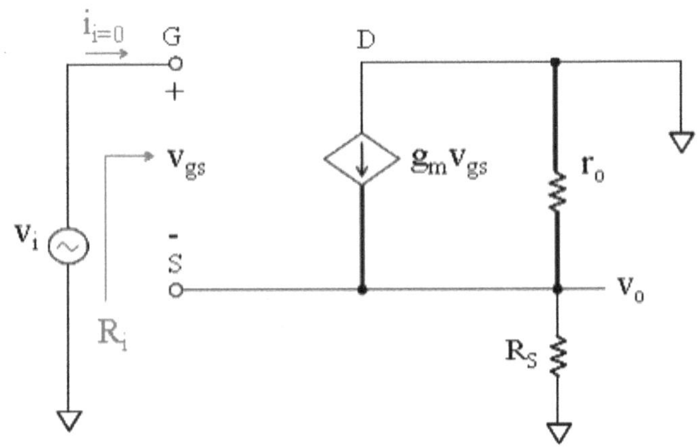


Fig. 17. ac equivalent circuit with a small signal equivalent model applied to find the small signal input resistance $R_i$.

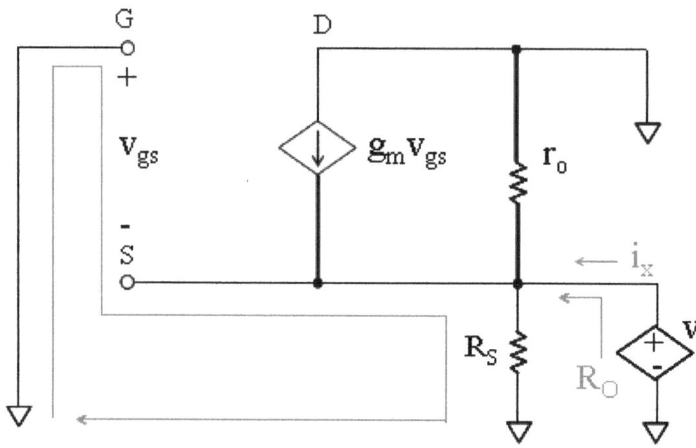


Fig. 18. ac equivalent circuit with a small-signal equivalent model applied to obtain the small signal output resistance $R_o$.

Fig. 18 shows an ac equivalent circuit using a small-signal equivalent model to find the small-signal output resistance $R_o$. When $v_i$=0, the current $i_x$ flowing when $v_x$ is applied to the output port is obtained as $R_o$.

As shown in Fig. 18, applying the KVL to the input loop yields Eq. (31), and if Eq. (31) is expanded for $v_{gs}$, Eq. (32) can be obtained. Eq. (33) can be obtained by applying the KCL to the source node.

$$v_{gs} + v_x = 0 \tag{31}$$

$$v_{gs} = - v_x \tag{32}$$

$$-i_x - g_m v_{gs} + \frac{v_x}{r_o} + \frac{v_x}{R_S} = 0 \tag{33}$$

In Eq. (33), $r_o$ and $R_S$ are parallel resistors and the resistance can be rewritten as in Eq. (34).

$$-i_x - g_m v_{gs} + \frac{v_x}{r_o||R_S} = 0 \tag{34}$$

Substituting Eq. (32) into Eq. (34) yields Eq. (35).

$$i_x = \left(g_m + \frac{1}{r_o||R_S}\right)v_x \tag{35}$$

So the small-signal output resistance $R_o$ is calculated as $v_x/i_x$ as shown in Eq. (36). If $g_m$ is larger than $1/(r_o \parallel R_s)$, $R_o$ is approximately equal to $1/g_m$.

$$R_o = \frac{v_x}{i_x} = \frac{1}{g_m + \frac{1}{r_o||R_S}} \cong \frac{1}{g_m} = r_s \tag{36}$$

## 5.5 Common Gate Amplifier

In the common gate amplifier circuit of Fig. 19, the gate node is connected to the DC supply voltage VGG, the input signal is connected to the source node, and the output signal is output from the drain node. A resistor $R_D$ is connected between the drain node of the MOS transistor and VDD.

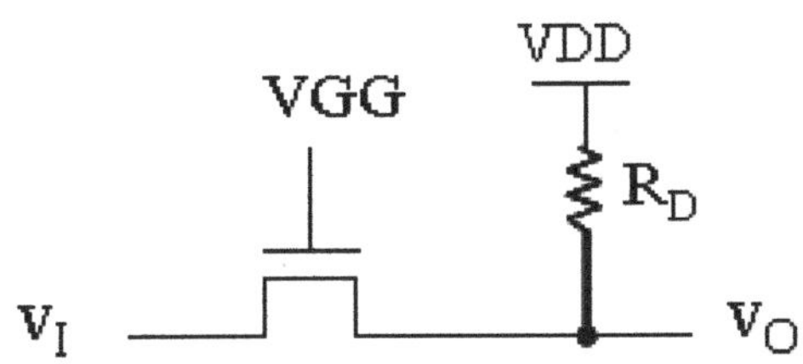


Fig. 19. Common gate amplifier.

The ac equivalent circuit of the common gate amplifier circuit can be obtained by shorting the DC voltage sources $V_I$, VGG, and VDD in the circuit of Fig. 19, as shown in Fig. 20.

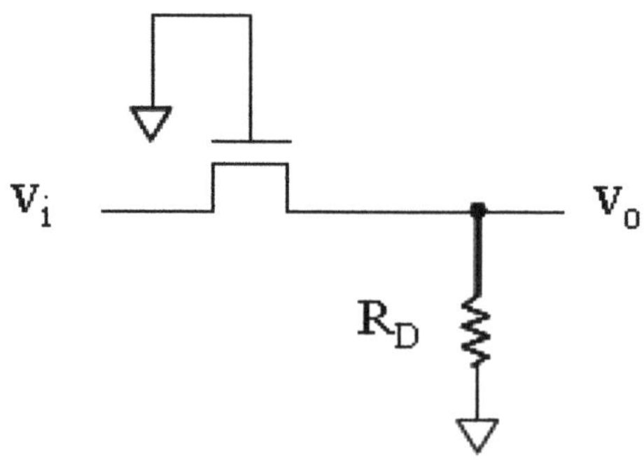


Fig. 20. ac equivalent circuit of a common gate amplifier.

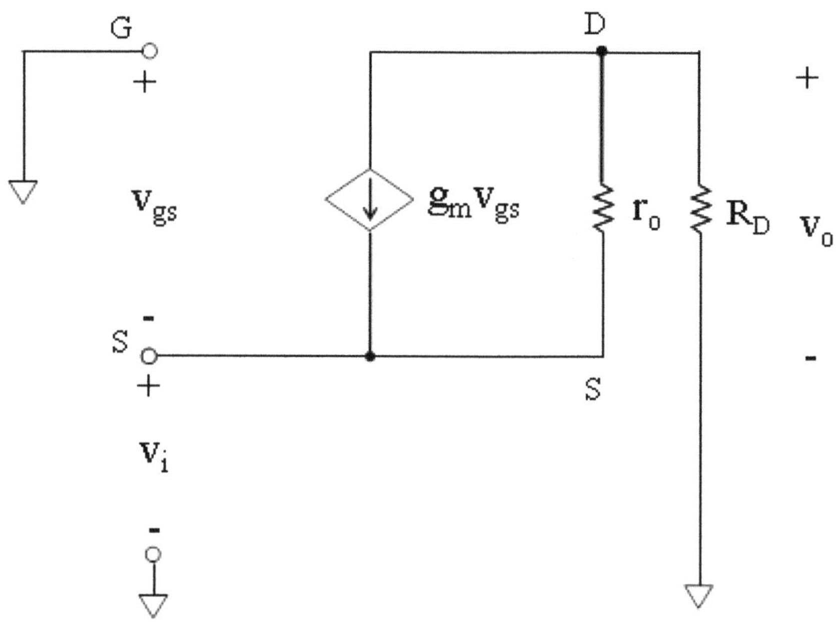


Fig. 21. ac equivalent circuit of a common gate amplifier circuit with a small-signal equivalent model applied.

Fig. 21 shows the ac equivalent circuit using the small-signal equivalent model, which is a hybrid-$\pi$ model of NMOS transistor in the ac equivalent circuit of Fig. 20. Applying the KVL to the input loop shown in Fig. 21 yields Eq. (37).

$$v_{gs} = -v_i \tag{37}$$

Then, applying the KCL to the source node in Fig. 21, we can obtain Eq. (38).

$$-g_m v_i + \frac{v_o - v_i}{r_o} + \frac{v_o}{R_D} = 0 \tag{38}$$

In Eq. (38), $r_o$ and $R_D$ are parallel resistors and the resistance can be rewritten as in Eq. (39).

$$\frac{v_o}{r_o \| R_D} = \left(g_m + \frac{1}{r_o}\right) v_i \tag{39}$$

Therefore, the voltage gain $A_v$ is given by Eq. (40).

$$A_v = \frac{v_o}{v_i} = \left(g_m + \frac{1}{r_o}\right)(r_o \| R_D) \cong g_m (r_o \| R_D) \tag{40}$$

Fig. 22 shows an ac equivalent circuit for finding the small-signal input resistance $R_i$. It is a resistor that is seen from the input port when $R_L$ is infinite.

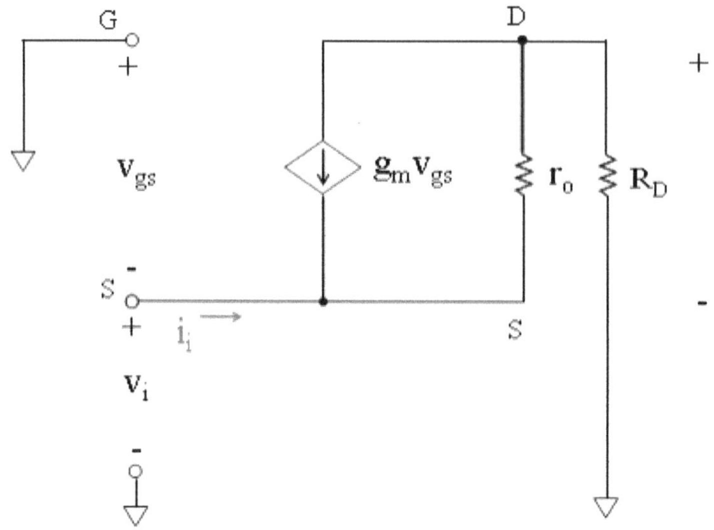


Fig. 22. ac equivalent circuit with a small signal equivalent model applied to find the small signal input resistance $R_i$.

Applying the KCL to the source node in Fig. 22 yields Eq. (41).

$$-i_i + g_m v_i + \frac{v_i - v_o}{r_o} = 0 \tag{41}$$

Then, by applying Ohm's law to $R_D$, we can obtain Eq. (42).

$$v_o = R_D i_i \tag{42}$$

Substituting Eq. (42) into Eq. (41) yields Eq. (43).

$$-\left(1+\frac{R_D}{r_o}\right)i_i + \left(g_m + \frac{1}{r_o}\right)v_i \cong -\left(1+\frac{R_D}{r_o}\right)i_i + g_m v_i = 0 \tag{43}$$

So $R_i$ is equal to Eq. (44).

$$R_i = \frac{v_i}{i_i}|_{R_L=\infty} = \frac{1}{g_m}\left(1+\frac{R_D}{r_o}\right) \tag{44}$$

If $r_o \gg R_D$, then $R_i$ is equal to Eq. (45).

$$R_i \cong \frac{1}{g_m} = r_s \tag{45}$$

Fig. 23 shows an ac equivalent circuit using a small-signal equivalent model to find the small-signal output resistance $R_o$. When $v_i$=0, the current $i_x$ flowing when $v_x$ is applied to the output port can be obtained as $R_o$.

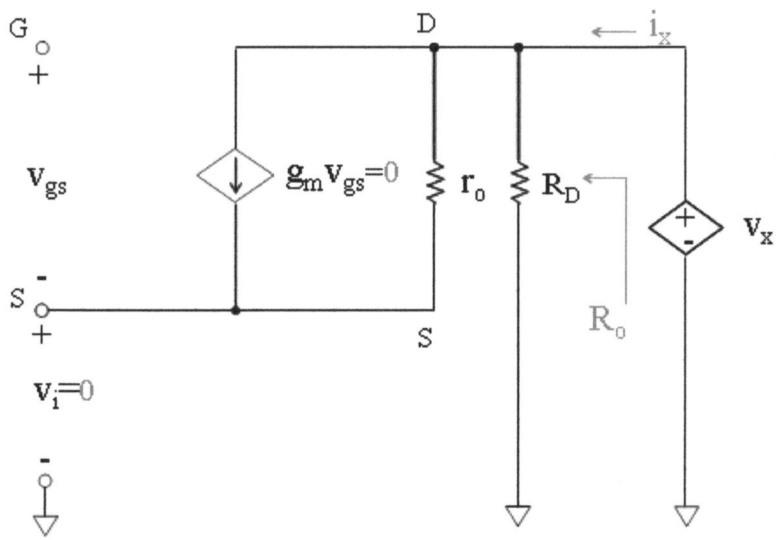


Fig. 23. ac equivalent circuit with a small-signal equivalent model applied to obtain the small signal input resistance $R_o$.

As shown in Fig. 23, the dependent current source $g_m v_{gs}$ is also zero because $v_{gs}$=$v_i$=0. Thus, the output resistance $R_o$ seen from the output port becomes $r_o \parallel R_D$ as in Eq. (46).

$$R_o = \frac{v_x}{i_x}\Big|_{v_i=0} = r_o \| R_D \tag{46}$$

Table 2 shows the characteristics of the common source, common gate, and common drain amplifier, which are single-stage MOS amplifiers.

Table 2. Comparison of characteristics of single-stage MOS amplifiers.

| | CS Amplifier | CG Amplifier | CD Amplifier |
|---|---|---|---|
| $A_v$ | $-g_m(r_o \parallel R_D)$ | $g_m(r_o \parallel R_D)$ | 1 |
| $R_i$ | $\infty$ | $\frac{1}{g_m}$ | $\infty$ |
| $R_o$ | $r_o \parallel R_D$ | $r_o \parallel R_D$ | $\frac{1}{g_m}$ |

**Example 5-1**

*Given that $R_D$=20 kΩ, $V_O$=2.04 V, Early voltage $V_A$=50 V, $V_{TN}$=0.7 V, $\mu_n C_{ox}$=70 μA/V², W=30 μm, and L=2 μm, answer the questions below when the amplifier circuit shown in Fig. 23 operates in the saturation region.*

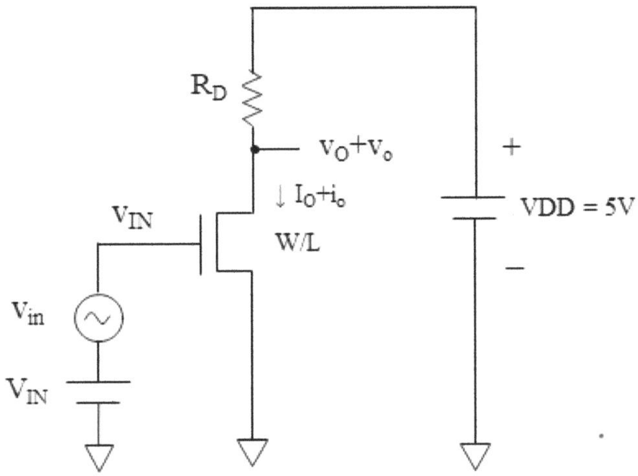


Fig. 23. An amplifier for Example 5-1.

(a) What kind of amplifier is this?

Question (a) relates to the type of amplifier, which is a common source amplifier since the source node of the NMOS transistor is connected to ac GND.

(b) Calculate $I_o$.

Question (b) is a problem of calculating the current $I_O$ at the DC operating point. Since the voltage across the drain resistor $R_D$ is $VDD - V_O$, dividing it by the resistance $R_D$ can calculate the current $I_O$ as shown in Eq. (47).

$$I_O = \frac{VDD - V_O}{R_D} \quad (47)$$

$$= \frac{5\ V - 2.04\ V}{20\ k\Omega}$$

$$= 148\ \mu A$$

(c) Calculate $g_m$.

Question (c) is a problem of calculating small-signal transconductance $g_m$, which can be obtained as in Eq. (48).

$$\begin{aligned} g_m &= \sqrt{2\mu_n C_{ox}\frac{W}{L}I_D} \\ &= \sqrt{2\times 70\mu A/V^2 \times \frac{30\ \mu m}{2\ \mu m}\times 148\ \mu A} \\ &= 0.56\ mS \end{aligned} \tag{48}$$

(d) Calculate $r_o$.

Question (d) is a problem of calculating the small-signal output resistance $r_o$. Before calculating $r_o$, the channel length modulation parameter $\lambda$ must be obtained using Early voltage as shown in Eq. (49).

$$\lambda = \frac{1}{V_A} = \frac{1}{50\ V} = 0.02\ V^{-1} \tag{49}$$

After $\lambda$ is obtained as in Eq. (49), small-signal output resistance $r_o$ can be calculated as in Eq. (50).

$$\begin{aligned} r_o &= \frac{1}{\lambda I_D} \\ &= \frac{1}{0.02\ V^{-1}\times 148\ \mu A} \\ &= 338\ k\Omega \end{aligned} \tag{50}$$

**Example 5-2**

*Given that* $V_{IN}$=*0.75 V,* $V_{TN}$=*0.5 V,* $\mu_n C_{ox}$=*70 μA/V², W=100 μm, and L=2 μm,* answer the following questions for the amplifier shown in Fig. 24.

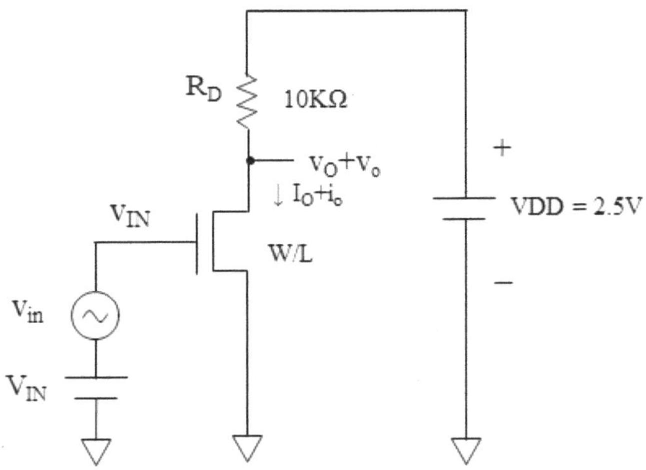


Fig 24. An amplifier for Example 5-2.

(a) Calculate $I_O$ and $V_O$ at the DC quiescent point.

Question (a) is a problem of calculating the current $I_O$ and the voltage $V_O$ at the DC operating point. Since $V_{GS}$ of the NMOS transistor is 0.75 V, which is the $V_{IN}$ voltage, the drain current $I_O$ can be calculated if we substitute 0.75 V into $V_{GS}$ as shown in Eq. (51).

$$I_O = \frac{1}{2}\mu_n C_{ox}\frac{W}{L}(V_{GS} - V_{TN})^2 \tag{51}$$

$$= \frac{1}{2}\times 70\mu A/V^2 \times \frac{100\mu m}{2\mu m}\times(0.75\,V - 0.5\,V)^2$$

$$= 109.4\mu A$$

In question (a), $V_O$ can be obtained by subtracting $I_O R_D$ voltage from $VDD$, as shown in Eq. (52).

$$\begin{aligned} V_O &= VDD - I_O R_D \\ &= 2.5\ V - 109.4\ \mu A \times 10\ k\Omega \\ &= 1.41\ V \end{aligned} \tag{52}$$

**(b) What is the drain voltage to be in the pinch-off region when the source voltage of the NMOS transistor is 0 V and the gate voltage is 0.75 V?**

Question (b) is a problem of calculating the drain voltage to be in the pinch-off region when the source voltage of the NMOS transistor is 0 V and the gate voltage is 0.75 V. The drain voltage $V_D$ is 0.25 V because the $V_{GD}$ voltage needs to be equal to $V_{TN}$ in order for the NMOS transistor to be in the pinch-off region.

**(c) Draw the $i_O - v_O$ curve of the NMOS transistor when the amplifier circuit of Fig. 24 is in DC state. Indicate the saturation current $I_O$. Channel length modulation effect is ignored.**

Question (c) presents the $i_O - v_O$ curve of the NMOS transistor and the saturation current $I_O$ when $V_{GS}$ is $V_{IN}$ in the DC condition of the amplifier circuit. The channel length modulation effect can be ignored. The $i_O - v_O$ curve of the NMOS transistor is shown in Fig. 25, and the saturation current $I_O$ is shown on the $i_O - v_O$ curve.

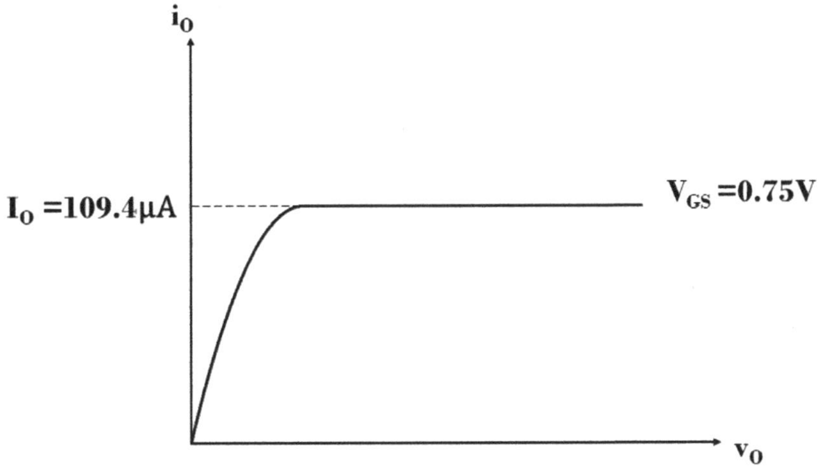


Fig. 25. The $i_O - v_O$ curve of the NMOS transistor when the amplifier circuit of Fig. 24 is in DC state.

(d) Draw the load line based on the result of (c).

Question (d) is a problem of drawing a load line on the $i_O - v_O$ curve in example (c). Load line $I_O$ is the same as Eq. (53). The slope is $-1/R_D$ and when $V_O$ is 0 V, $I_O$ is $VDD/R_D$.

$$I_O = \frac{VDD - V_O}{R_D} \tag{53}$$

As shown in Fig. 26, the operating point is where the $i_O - v_O$ curve meets the load line in the DC state.

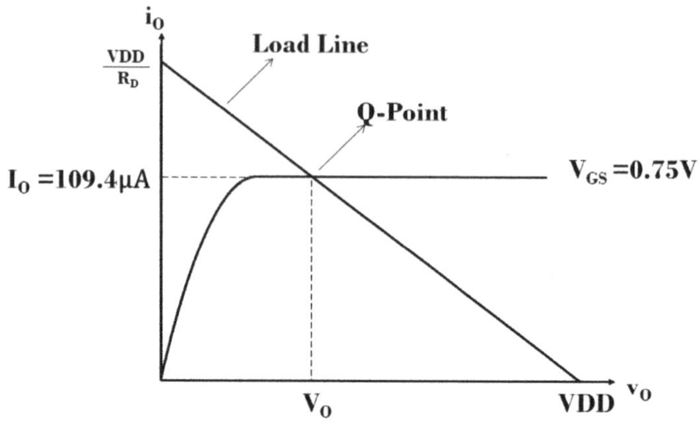


Fig. 26. The $i_O - v_O$ curve and load line of an NMOS transistor.

### Example 5-3

*Given that VDD=5 V, $R_D$=20 kΩ, $\mu_n C_{ox}$=70 μA/V², $V_{TN}$=0.7 V, W=100 μm, L=2 μm,* $V_A$=50 V, and $I_O$=$I_D$=70 μA, answer the following questions for the amplifier shown in Fig. 27.

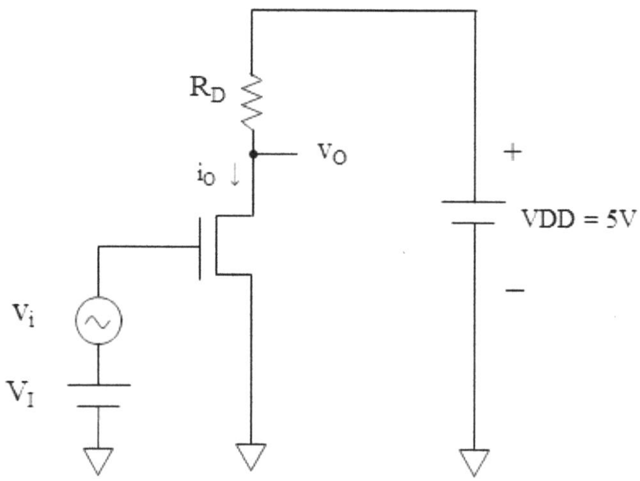


Fig. 27. An amplifier for Example 5-3.

(a) Calculate $V_I$ and $V_O$ at the DC quiescent point. Here, $v_i$=0.

Question (a) is to calculate $V_I$ and $V_O$ at the operating point. Since $I_O$ is the drain current $I_D$ and $V_{GS}$ is $V_I$, the equation for $I_O$ can be written as Eq. (54).

$$I_O = \frac{1}{2}\mu_n C_{ox}\frac{W}{L}(V_I - V_{TN})^2 \tag{54}$$

Substituting each parameter into Eq. (54), we can write Eq. (55).

$$70\ \mu A = \frac{1}{2}\times 70\ \mu A/V^2 \times \frac{100\ \mu m}{2\ \mu m}(V_I - 0.7)^2 \tag{55}$$

And if you expand Eq. (55) for $V_I$, the $V_I$ is obtained as in Eq. (56).

$$V_I = 0.9\ V \tag{56}$$

In addition, $V_O$ can be obtained as shown in Eq. (57).

$$\begin{aligned} V_O &= VDD - I_O R_D \\ &= 5\ V - 70\ \mu A \times 20\ k\Omega \\ &= 3.6\ V \end{aligned} \tag{57}$$

**(b) Calculate small-signal parameters $g_m$, $r_o$, and $r_s$.**

Question (b) is to calculate the small-signal parameters $g_m$, $r_o$, and $r_s$. Small-signal transconductance $g_m$ can be obtained as shown in Eq. (58).

$$\begin{aligned} g_m &= \sqrt{2\mu_n c_{ox}\frac{W}{L}I_D} \\ &= \sqrt{2\times 70\mu A/V^2\times\frac{100\ \mu m}{2\ \mu m}\times 70\ \mu A} \\ &= 0.7\ mS \end{aligned} \tag{58}$$

To obtain the small-signal output resistance $r_o$, the channel length modulation parameter $\lambda$ is first obtained from the Early voltage as shown in Eq. (59).

$$\lambda = \frac{1}{V_A} = \frac{1}{50\ V} = 0.02\,V^{-1} \tag{59}$$

Substituting $\lambda$ obtained from Eq. (59) into Eq. (60) yields the small-signal output resistance $r_o$.

$$\begin{aligned} r_o &= \frac{1}{\lambda I_D} \\ &= \frac{1}{0.02\,V^{-1}\times 70\ \mu A} \\ &\cong 714.286\ k\Omega \end{aligned} \tag{60}$$

In addition, small-signal source resistance $r_s$ can be obtained using $g_m$ as in Eq. (61).

$$\begin{aligned} r_s &= \frac{1}{g_m} \\ &= \frac{1}{0.7\,mS} \\ &\cong 1.429\,k\Omega \end{aligned} \tag{61}$$

(c) Draw an ac equivalent circuit.

Question (c) is the problem of drawing an ac equivalent circuit. The ac equivalent circuit can be drawn like the circuit shown in Fig. 28 by shortening the DC voltage sources $V_I$ and VDD in the given amplifier circuit.

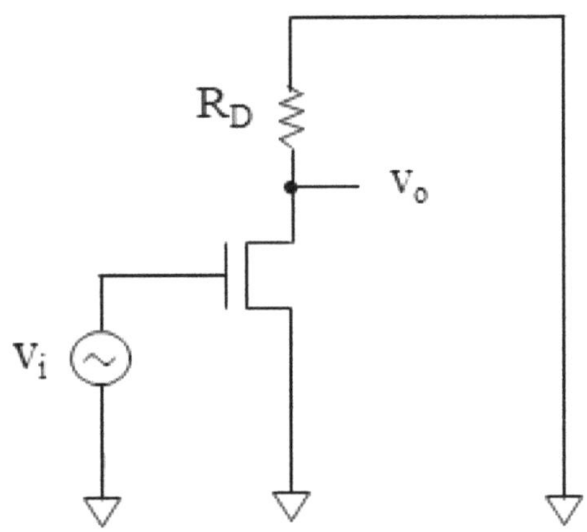


Fig. 28. ac equivalent circuit.

**(d) Draw the small-signal equivalent circuit based on the hybrid-$\pi$ model.**

Question (d) is a problem of drawing small-signal equivalent circuit applying hybrid-$\pi$ model. If we replace hybrid-$\pi$ model instead of NMOS transistor, the equivalent circuit can be drawn as in Fig. 29.

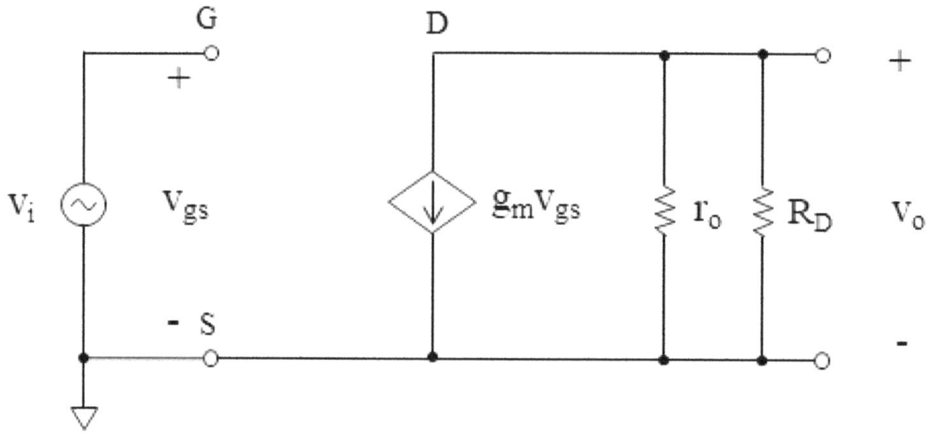


Fig. 29. Small-signal equivalent circuit based on the hybrid-π model.

**(e) Calculate the voltage gain $A_V$.**

Question (e) is a problem of calculating the voltage gain $A_v$. First, if the KVL is applied to the output port of the small signal equivalent circuit using the hybrid-$\pi$ model, $v_o$ is as shown in Eq. (62).

$$\begin{aligned} v_o &= -g_m v_{gs} \times (r_o \| R_D) \\ &= -g_m v_i \times (r_o \| R_D) \end{aligned} \tag{62}$$

Since the voltage gain $A_v$ is $v_o/v_i$, Eq. (63) can be obtained by rearranging Eq. (62) for $v_o/v_i$.

$$\begin{aligned} A_V &= \frac{v_o}{v_i} \\ &= -g_m(r_o \| R_D) \\ &= -0.7\ mS \times (\frac{714.286\ k\Omega \times 20\ k\Omega}{714.286\ k\Omega + 20\ k\Omega}) \\ &= -13.619 \end{aligned} \tag{63}$$

(f) Calculate the input resistance $R_i$.

Question (f) is the problem of calculating small-signal input resistance $R_i$. As shown in Fig. 30, in the ac equivalent circuit when the hybrid $\pi$-model is applied, ac input current $i_i$ is 0, and so $R_i$ is infinite as shown in Eq. (64).

$$\begin{aligned} R_i &= \frac{v_i}{i_i}\Big|_{R_L=\infty} \\ &= \frac{v_i}{0} \\ &= \infty \end{aligned} \tag{64}$$

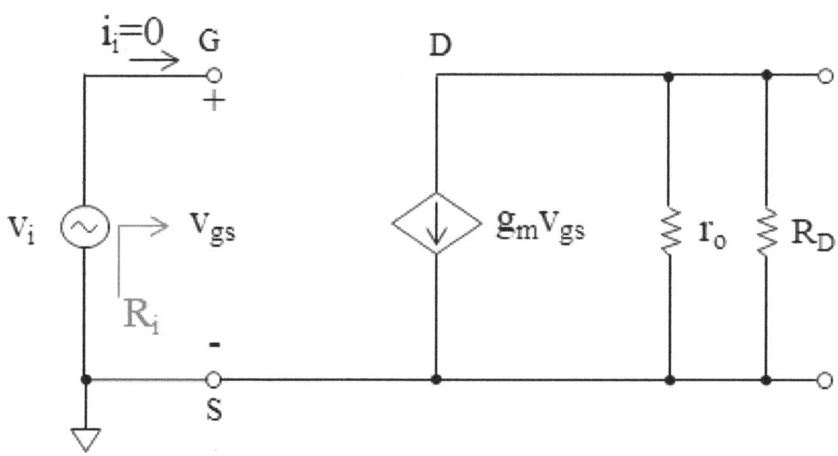


Fig. 30. ac equivalent circuit using hybrid π-model to obtain $R_i$.

(g) Calculate the output resistance $R_o$.

Question (g) is a problem of calculating small-signal output resistance $R_o$. $R_o$ is the equivalent circuit of Fig. 31 because it is the resistance seen from the output port when $v_i$ is zero. Applying the KVL at the output port, $v_x$ is equal to Eq. (65).

$$v_x = i_x (r_o || R_D) \tag{65}$$

$R_o$ can be calculated as shown in Eq. (66).

$$\begin{aligned} R_o &= \frac{v_x}{i_x} \\ &= \frac{i_x (r_o || R_D)}{i_x} \\ &= r_o || R_D \\ &= \frac{714.286\ k\Omega \times 20\ k\Omega}{714.286\ k\Omega + 20\ k\Omega} \\ &= 19.46\ k\Omega \end{aligned} \tag{66}$$

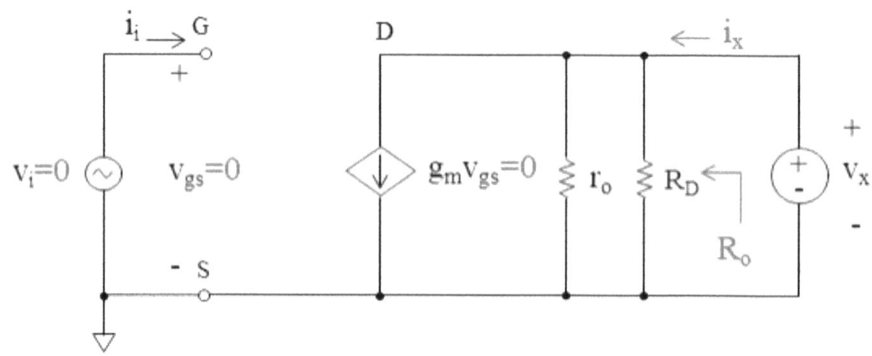


Fig. 31. ac equivalent circuit using hybrid π-model to obtain $R_o$.

**Example 5-4**

*Given that VDD=5 V, $R_S$=10 kΩ, $\mu_n C_{ox}$=70 μA/V², $V_{TN}$=0.7 V, W=100 μm, L=2 μm, and $V_A$=50 V, answer the following questions for the amplifier shown in Fig. 32.*

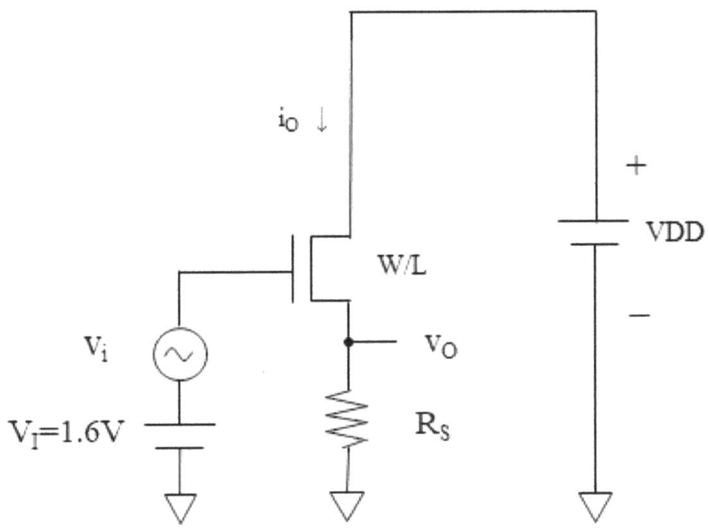


Fig. 32. An amplifier for Example 5-4.

(a) What kind of amplifier is this?

Question (a) relates to the type of amplifier shown in Fig. 32, which is a common drain amplifier because the drain node is connected to VDD, which is the ac GND.

(b) Calculate $I_O$ and $V_O$ at the DC quiescent point.

Question (b) is the problem of calculating $I_O$ and $V_O$ at DC operating point. When the Eq. (67), which is the current equation flowing through the resistor $R_S$, and the Eq. (68), which is the drain current of the NMOS transistor, are combined and solved, the solutions of $V_{GS}$ are 0.443 V and 0.9 V as shown in Eq. (69).

$$I_O = \frac{V_O}{R_S} = \frac{V_I - V_{GS}}{R_S} = \frac{1.6\ V - V_{GS}}{10\ k\Omega} \tag{67}$$

$$\begin{aligned} I_O &= \frac{1}{2}\mu_n C_{ox}\frac{W}{L}(V_{GS} - V_{TN})^2 \\ &= \frac{1}{2}\times 70\ \mu A/V^2 \times \frac{100\ \mu m}{2\ \mu m}\times (V_{GS} - 0.7\ V)^2 \end{aligned} \tag{68}$$

$$V_{GS} = 0.443\ V \text{ or } 0.9\ V \tag{69}$$

In the $V_{GS}$ solutions of 0.443 V and 0.9 V, since 0.443 V is less than 0.7 V, which is the $V_{TN}$, this condition does not meet the prerequisite that the NMOS transistors must be in the saturation region. So, the $V_{GS}$ solution is 0.9 V. In addition, the $V_O$ voltage can be obtained by subtracting the $V_{GS}$ voltage from $V_I$ as shown in Eq. (70).

$$\begin{aligned} V_O &= V_I - V_{GS} \\ &= 1.6\ V - 0.9\ V \\ &= 0.7\ V \end{aligned} \tag{70}$$

Furthermore, the $I_O$ current is 70 μA when calculated as shown in Eq. (71).

$$I_O = \frac{V_O}{R_s} = \frac{0.7\ V}{10\ K\Omega} = 70\ \mu A \tag{71}$$

(c) Calculate small-signal parameters $g_m$ and $r_o$.

Question (c) is the problem of calculating the small-signal parameters $g_m$ and $r_o$. $g_m$ is calculated as in Eq. (72), which is 0.7 mS.

$$\begin{aligned} g_m &= \sqrt{2\mu_n C_{ox}\frac{W}{L}I_{DS}} \\ &= \sqrt{2\times 70\ \mu A/V^2\times\frac{100\ \mu m}{2\ \mu m}\times 70\ \mu A} \\ &= 0.7\ mS \end{aligned} \tag{72}$$

In order to calculate $r_o$, the channel length modulation parameter $\lambda$ is first calculated using Early voltage as shown in Eq. (73).

$$\lambda = \frac{1}{V_A} = \frac{1}{50\ V} = 0.02\ V^{-1} \tag{73}$$

$r_o$ can be calculated by substituting $\lambda$ calculated in equation (73) into Eq. (74).

$$\begin{aligned} r_o &= \frac{1}{\lambda I_D} \\ &= \frac{1}{0.02\ V^{-1}\times 70\ \mu A} \\ &= 714.286\ k\Omega \end{aligned} \tag{74}$$

(d) Draw the ac equivalent circuit.

Question (d) is the problem of drawing an ac equivalent circuit for a common drain amplifier. To draw an ac equivalent circuit, $V_I$ and VDD are shorted, and the circuit is shown in Fig. 33.

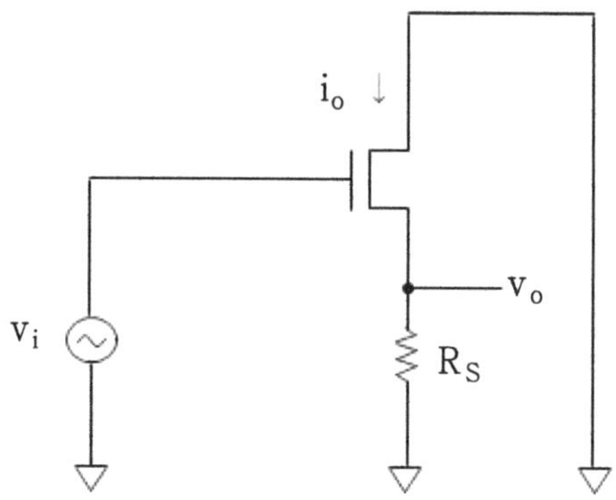


Fig. 33. ac equivalent circuit.

(e) Draw the small-signal equivalent circuit based on the hybrid-$\pi$ model.

Question (e) is a problem of drawing ac equivalent circuit applying hybrid $\pi$-model. Replacing the hybrid $\pi$-model instead of an NMOS transistor allows us to draw an ac equivalent circuit as shown in Fig. 34.

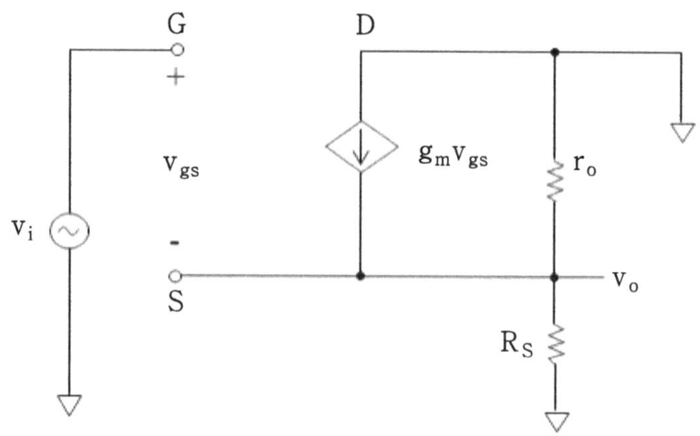


Fig. 34. ac equivalent circuit.

(f) Calculate the voltage gain $A_V$.

Question (f) is the problem of calculating the voltage gain. Applying the KVL to the input loop first gives Eq. (75).

$$-v_i + v_{gs} + v_o = 0 \tag{75}$$

Applying the KCL to the source node gives Eq. (76).

$$-g_m v_{gs} + \frac{v_o}{R_S} + \frac{v_o}{r_o} = 0 \tag{76}$$

Since $R_S$ and $r_o$ are parallel connections, Eq. (76) can be written as Eq. (77).

$$-g_m v_{gs} + \frac{v_o}{r_o \| R_s} = 0 \tag{77}$$

When Eq. (75) and Eq. (76) are combined and solved, Eq. (78) can be obtained.

$$v_i = \left(1 + \frac{1}{g_m (r_o \| R_s)}\right) \times v_o \tag{78}$$

Since the voltage gain $A_v$ is $v_o/v_i$, the Eq. (78) can be rearranged with respect to $v_o/v_i$ as in Eq. (79).

$$\begin{aligned} A_v &= \frac{v_o}{v_i} \\ &= \frac{1}{1 + \frac{1}{g_m (r_o \| R_S)}} \\ &= 0.873 \end{aligned} \tag{79}$$

(g) Calculate the input resistance $R_i$.

Question (g) is the problem of calculating small-signal input resistance $R_i$. $R_i$ is the resistance seen from the input port when the load resistance $R_L$ is infinite as shown in Fig. 35, and $R_i$ is infinite because the input current $i_i$ is zero.

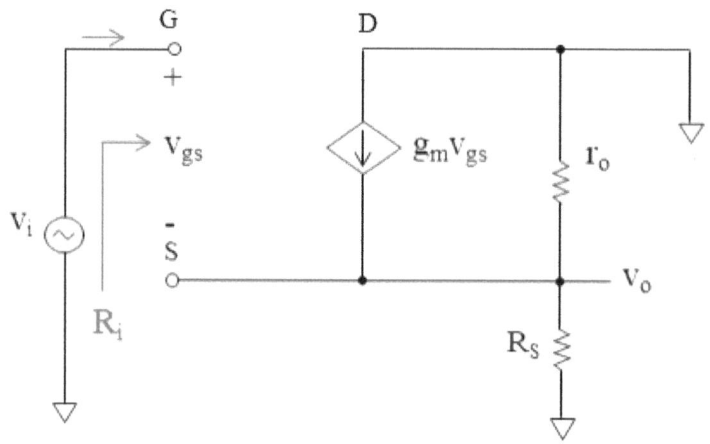


Fig. 35. ac equivalent circuit using hybrid π-model to obtain $R_i$.

(h) Calculate the output resistance $R_o$.

Question (h) is the problem of calculating small-signal output resistance $R_o$. In order to calculate $R_o$, applying the KVL to the input loop shown in Fig. 36, Eq. (80) can be obtained.

$$v_{gs} + v_x = 0 \tag{80}$$

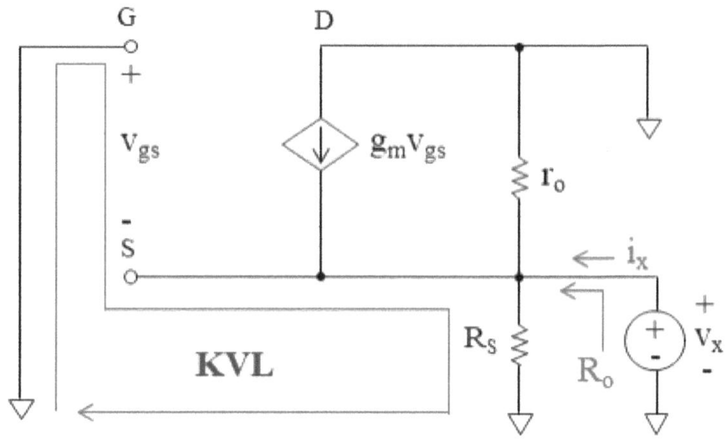


Fig. 36. ac equivalent circuit using hybrid π-model to obtain $R_o$.

And by applying the KCL to the source node, Eq. (81) can be obtained.

$$-i_x - g_m v_{gs} + \frac{v_x}{r_o \| R_S} = 0 \tag{81}$$

When Eqs. (80) and (81) are combined and solved, Eq. (82) is obtained.

$$i_x = \left(g_m + \frac{1}{r_o \| R_S}\right) v_x \tag{82}$$

Since $R_o$ is $v_x / i_x$, if Eq. (82) is rewritten for $v_x / i_x$, $R_o$ is equal to Eq. (83), and $R_o$ is 1.248 kΩ.

$$\begin{aligned} R_o &= \frac{v_x}{i_x} \\ &= \frac{1}{g_m + \frac{1}{r_o \| R_S}} \\ &= 1.248\ k\Omega \end{aligned} \tag{83}$$

## Exercises

1. Answer the following questions.

   (a) Write down three types of 1-stage amplifiers that use MOS transistors..

   (b) Draw a hybrid-$\pi$ model of NMOS transistor.

   (c) Draw a T-model of NMOS transistor.

2. Please answer the following questions about the amplifier circuit below.

Here, $V_{TN}$=0.5 V, $\mu_n C_{ox}$ = 70 $\mu$A/V$^2$, $W$=10 $\mu$m, $L$=2 $\mu$m, $M$=10, and $V_A$=50 V

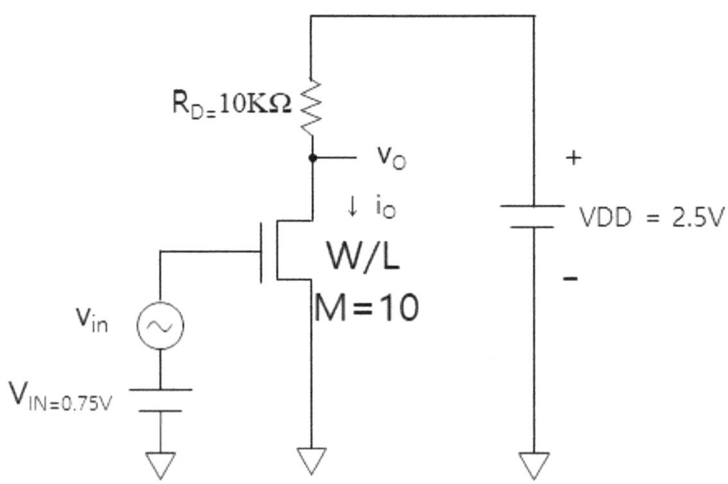

(a) Calculate $I_O$ and $V_O$ at DC operating point. $v_{in}$=0.

(b) Calculate small-signal parameters $g_m$, $r_o$, and $r_s$.

(c) Draw ac equivalent circuit with hybrid-$\pi$ model applied.

(d) Calculate voltage gain $A_v$.

(e) Calculate input resistance $R_i$.

$$R_i = \frac{v_i}{i_i}\Big|_{R_L=\infty}$$

(f) Calculate output resistance $R_O$.

$$R_o = \frac{v_x}{i_x}\Big|_{v_i=0}$$

3. Answer the following questions about the amplifier circuit below.
Here, $V_{TN}$=0.5 V, $\mu_n C_{ox}$=70 μA/V$^2$, W=25 μm, L=2 μm, $V_A$=50 V, and $R_S$=10 kΩ.

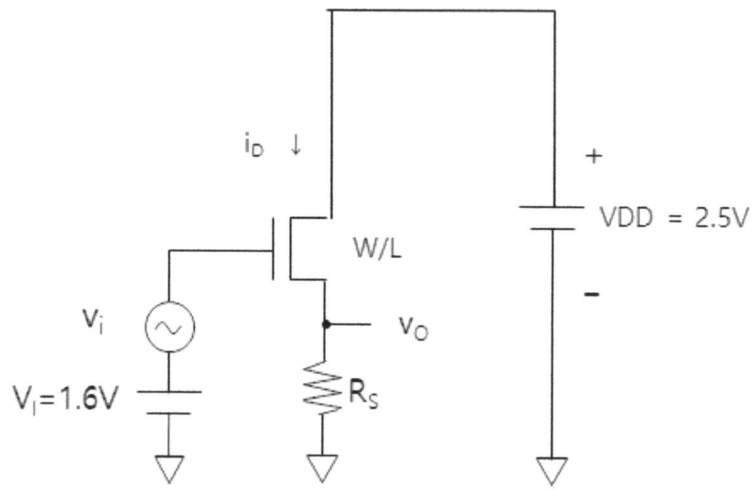


(a) Calculate $I_D$ and $V_O$ at DC operating point. $v_{in}$ =0.

(b) Calculate small-signal parameters $g_m$, $r_o$, and $r_s$.

(c) Draw a DC equivalent circuit.

(d) Draw an ac equivalent circuit with hybrid-$\pi$ model applied.

(e) Calculate voltage gain $A_v$.

(f) Calculate input resistance $R_i$.

$$R_i = \frac{v_i}{i_i}\Big|_{R_L = \infty}$$

(g) Calculate output resistance $R_o$.

$$R_o = \frac{v_x}{i_x}\Big|_{v_i=0}$$

4. Draw a common gate amplifier circuit.

5. Given that $\mu_n c_{ox}$=70 μA/V$^2$ , $V_{TN}$=0.5 V, W=100 μm, L=2 μm, $V_I$=0.7 V, VDD=2.5 V, $R_D$=15 kΩ, and $V_A$=50 V.

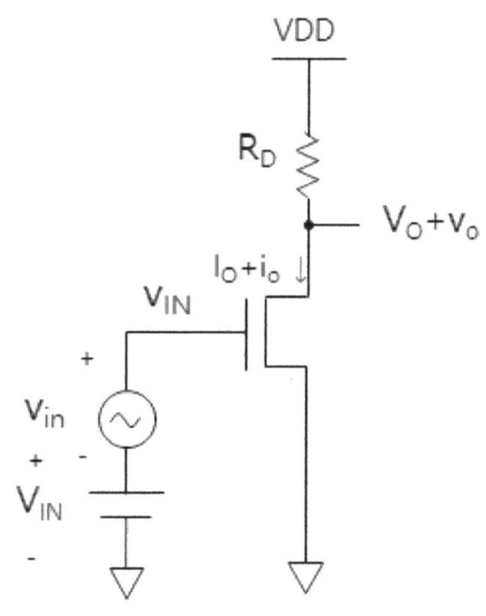


(a) Calculate $I_D$ and $V_O$ at DC operating point.

(b) Calculate small-signal parameters $g_m$ and $r_o$.

(c) Draw ac equivalent circuit with hybrid-$\pi$ model applied.

(d) Calculate voltage gain $A_v$.

(e) Calculate input resistance $R_i$.

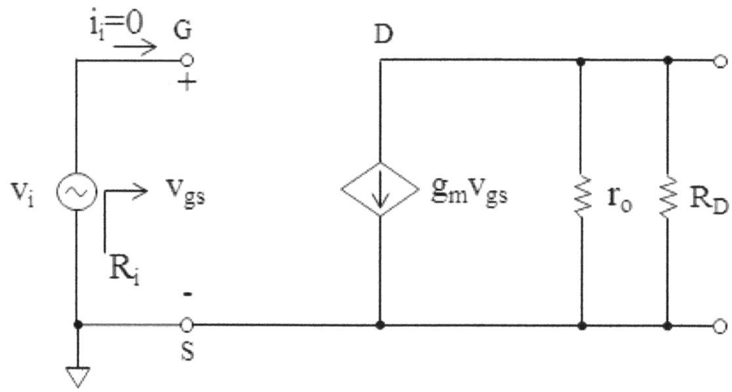


(f) Calculate output resistance $R_o$.

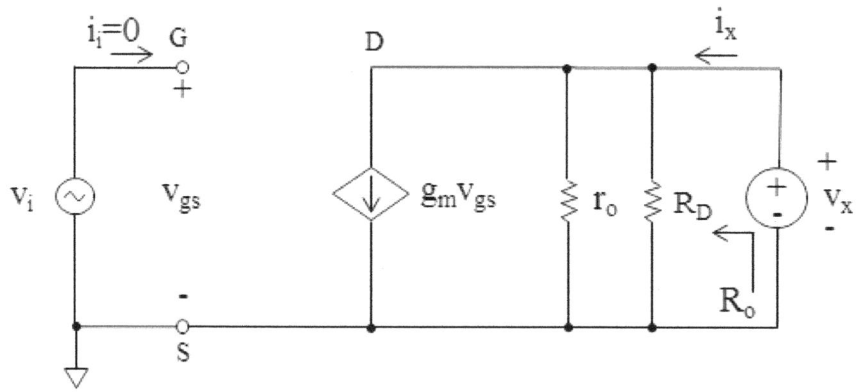

6. Answer the following questions.

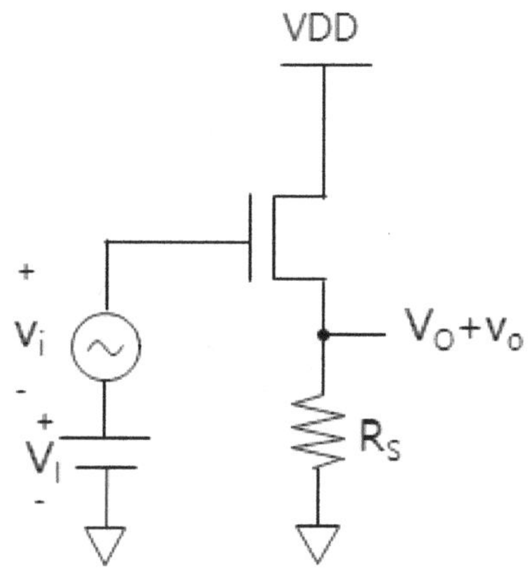


(a) What kind of amplifier circuit is this amplifier circuit?

(b) Draw a DC equivalent circuit.

(c) Draw an ac equivalent circuit with hybrid-$\pi$ model applied.

(d) Derive voltage gain $A_v$.

(e) Derive small-signal input resistance $R_i$.

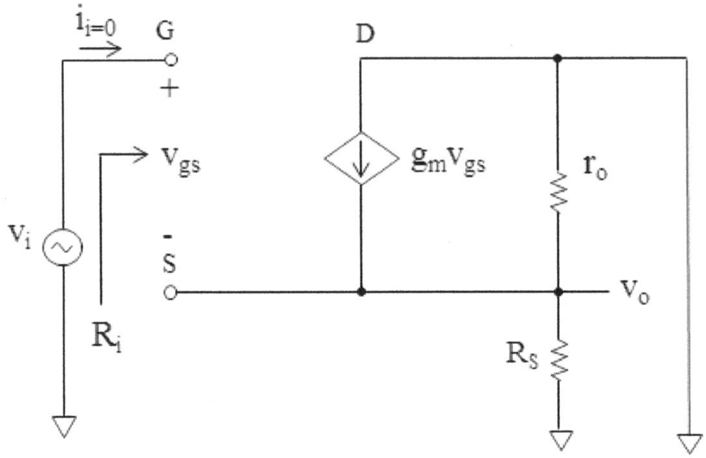


(f) Derive small-signal output resistance $R_o$.

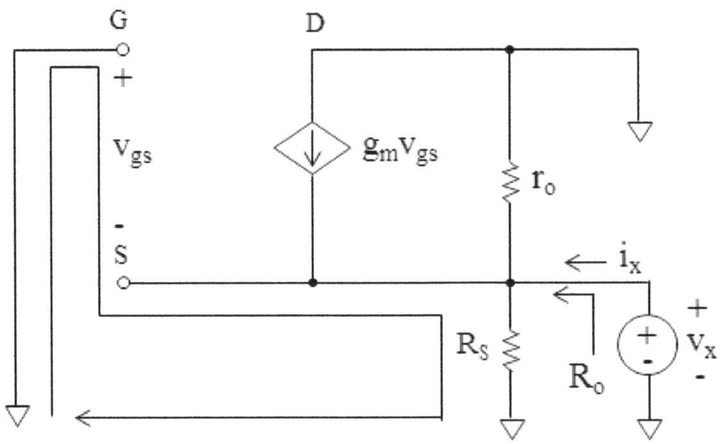

7. Answer the following questions about the amplifier circuit below.

VDD=5 V, $R_D$=20 KΩ, $\mu_n C_{ox}$=70 μA/V², $V_{TN}$=0.7 V, W=100 μm, L=2 μm, $V_A$=50 V, and $I_O$=$I_{DS}$=70 μA

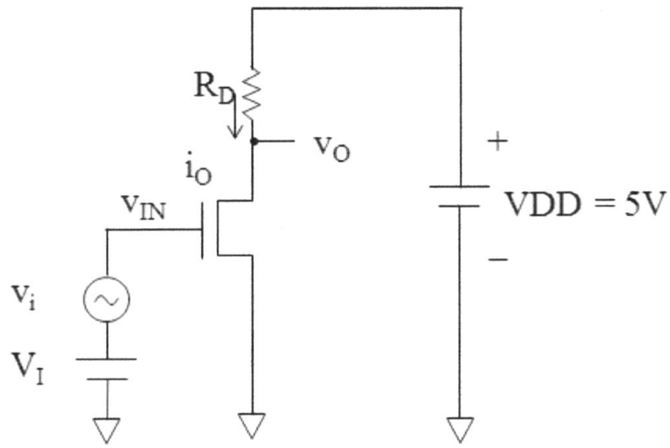


(a) What kind of amplifier circuit is this amplifier circuit?

(b) Calculate $I_O$ and $V_O$ at DC operating point. $v_{in}$ =0.

(c) Calculate small-signal parameters $g_m$, $r_o$, and $r_s$.

(d) Draw ac equivalent circuit.

(e) Draw ac equivalent circuit with hybrid-$\pi$ model applied.

(f) Calculate voltage gain $A_v$.

$$A_v = \frac{v_o}{v_i}$$

(g) Calculate input resistance $R_i$.

$$R_i = \frac{v_i}{i_i}\Big|_{R_L=\infty}$$

(h) Calculate output resistance $R_o$.

$$R_o = \frac{v_x}{i_x}\Big|_{v_i=0}$$

8. Answer the following questions about the amplifier circuit below.
Here, VDD=5 V, $R_S$=10 kΩ, $\mu_n c_{ox}$=70 μA/V$^2$, $V_{TN}$=0.7 V, W=100 μm, L=2 μm, and $V_A$ = 50 V.

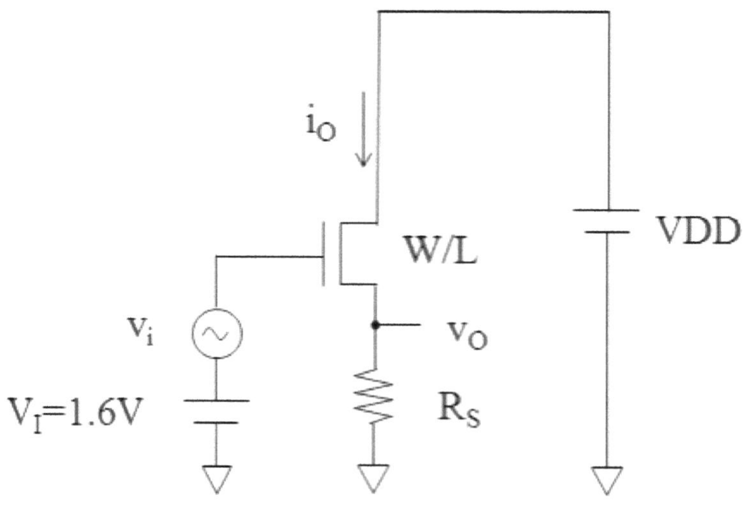

(a) What kind of amplifier circuit is this amplifier circuit?

(b) Calculate $I_O$ and $V_O$ at DC operating point. $v_{in}$ =0.

(c) Calculate small-signal parameters $g_m$, $r_o$, and $r_s$.

(d) Draw ac equivalent circuit.

(e) Draw ac equivalent circuit with hybrid-$\pi$ model applied.

(f) Calculate voltage gain $A_v$.

$$A_v = \frac{v_o}{v_i}$$

(g) Calculate input resistance $R_i$.

$$R_i = \frac{v_i}{i_i}\Big|_{R_L=\infty}$$

(h) Calculate output resistance $R_o$.

$$R_o = \frac{v_x}{i_x}\Big|_{v_i=0}$$

## Chap. 5 References

[1] Hongjun Park, Design of CMOS Analog Integrated Circuits, $\Sigma$Sigma Press, 1999.

[2] B. Razavi, Design of Analog CMOS Integrated Circuits, Mc-Graw Hill Education, 2017.

[3] Sedra et al., Microelectronic Circuits, Oxford.

CHAPTER 6

# Differential Amplifiers

The differential amplifier has characteristics that it amplifies the differential mode input voltage, which is the difference voltage between the two input voltages of $v_{I1}$ and $v_{I2}$ shown in Fig. 1, and outputs the amplified differential mode input voltage to the output voltage, but can not output the amplified common mode input voltage of the two input voltages to the output node since the voltage gain is low for the common mode input voltage. Differential amplifiers are most commonly used in analog integrated circuits. They are used for input stages of operational amplifiers, memory sense amplifiers, and voltage regulator comparators. There are two differential amplifiers depending on the types of load. One is a differential amplifier circuit having a resistive load as shown in Fig. 1, and the other is a differential amplifier circuit having an active load.

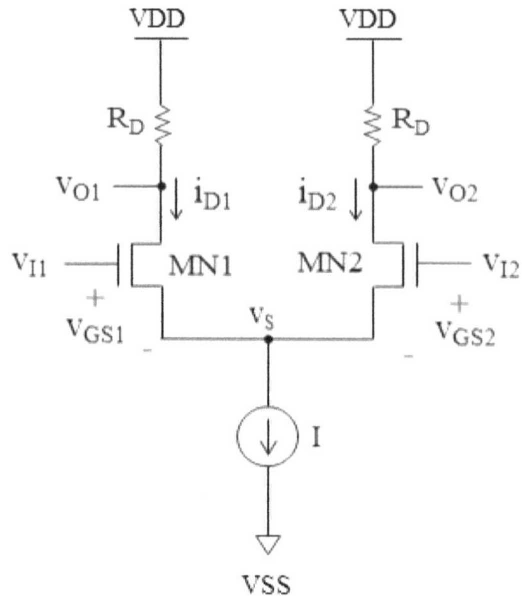


Fig. 1. Differential amplifier circuit with resistive load.

## 6.1 Differential Amplifier with Resistive Load

Fig. 1 shows a differential amplifier circuit with resistive load. The two identical NMOS transistors MN1 and MN2 constitute the input stage and the source nodes of MN1 and MN2 are connected together and an ideal constant current source I is connected between the source node and VSS. Resistors $R_D$ are connected between $v_{O1}$ and VDD, and $v_{O2}$ and VDD, respectively.

When the common mode input voltage $V_{CM}$ is simultaneously applied to $v_{G1}$ and $v_{G2}$ as shown in Fig. 2, $v_{G1}=v_{G2}=V_{CM}$. Assuming that the MN1 and MN2 transistors are identical and are in the saturation region, the currents $i_{D1}$ and $i_{D2}$ flowing through MN1 and MN2 are I/2, because the gate voltages $v_{G1}$ and $v_{G2}$ are equal to $V_{CM}$ and the source node voltages of MN1 and MN2 are the same.

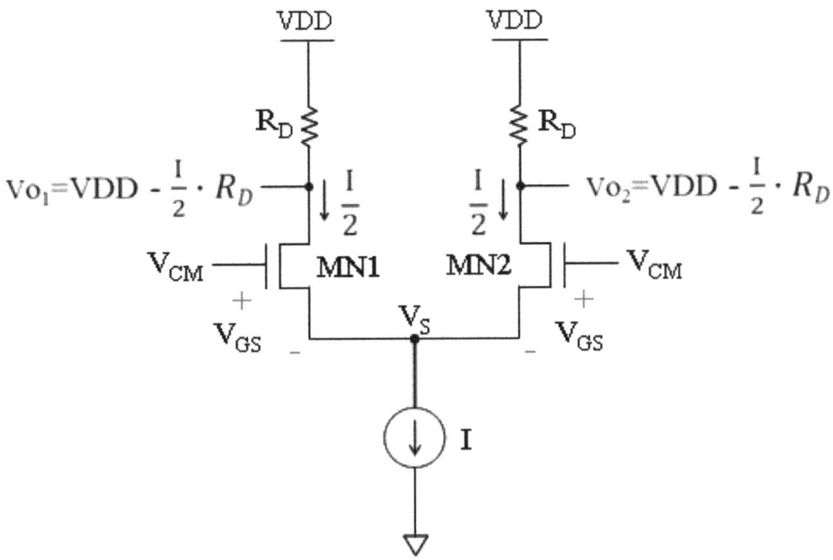


Fig. 2. In case that a common mode input voltage is applied to a differential amplifier having resistive load as load.

If the NMOS transistor is in the saturation region and channel length modulation is not considered, the current-voltage relationship is given by Eq. (1).

$$I_D = \frac{1}{2}\mu_n C_{ox}\frac{W}{L}(V_{GS} - V_{TN})^2 \tag{1}$$

The $V_{DSAT}$ voltage, which is the over-driving voltage for the NMOS transistor to be in the saturation region, can be written as Eq. (2).

$$V_{DSAT} = V_{GS} - V_{TN} = \sqrt{\frac{2I_D}{\mu_n C_{ox}\frac{W}{L}}} = \sqrt{\frac{I}{\mu_n C_{ox}\frac{W}{L}}} \tag{2}$$

On the other hand, the source node voltage $V_S$ in Fig. 2 equals to Eq. (3).

$$V_S = V_{CM} - V_{GS} = V_{CM} - (V_{TN} + V_{DSAT}) = V_{CM} - V_{TN} - \sqrt{\frac{I}{\mu_n C_{ox} \frac{W}{L}}} \tag{3}$$

The output voltages $v_{O1}$ and $v_{O2}$ of the differential amplifier are shown in Eqs. (4) and (5), and the output voltage difference $v_O(= v_{O1} - v_{O2})$ is zero. In this way, for the common mode input voltage $V_{CM}$, if MN1 and MN2 transistors in Fig. 2 are identical and operate in the saturation region, the current I of the tail current source is divided in half. If the load resistance $R_D$ connected to $v_{O1}$ and $v_{O2}$ are the same, the differential output voltage $v_O$ of the differential pair does not change. In other words, in a differential amplifier circuit with the same load resistance, the common mode input voltage signal does not affect the differential output voltage.

One of the important characteristics of the differential amplifier is the common mode input voltage range, which is the $V_{CM}$ voltage range in which the NMOS transistors MN1 and MN2 of the input differential pair operate in the saturation region.

$$v_{o1} = VDD - \frac{I}{2} R_D \tag{4}$$

$$v_{o2} = VDD - \frac{I}{2} R_D \tag{5}$$

In general, the current source of the NMOS differential pair uses the NMOS current mirror shown in Fig. 3. The NMOS current mirror circuit gets the $V_{BIAS}$ voltage through the MN3 transistor in the saturation region when a reference current $I_{REF}$ current flows through MN3 which is a diode-connected NMOS transistor. The $V_{BIAS}$ voltage can be obtained by using Eq. (6).

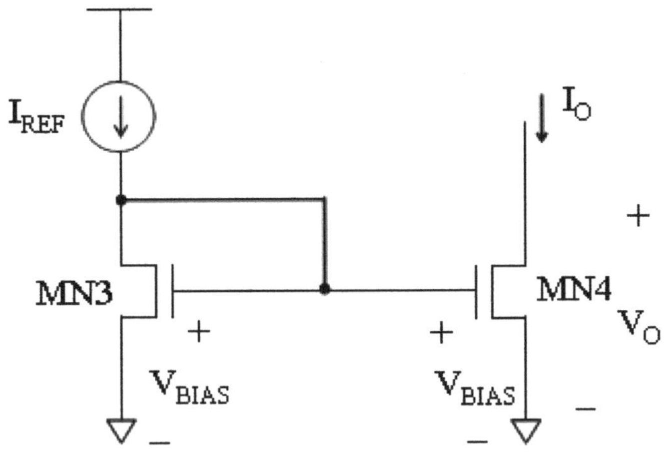


Fig. 3. NMOS current mirror circuit.

$$I_{REF} = \frac{1}{2}\mu_n Cox\frac{W}{L}(V_{BIAS} - V_{TN})^2 \tag{6}$$

$$V_{BIAS} = V_{TN} + V_{DSAT.MN3} = V_{TN} + \sqrt{\frac{2I_{REF}}{\mu_n C_{ox}\frac{W}{L}}} \tag{7}$$

As shown in Eq. (7), the $V_{BIAS}$ voltage is the sum of $V_{TN}$ and $V_{DSAT.MN3}$. The current flowing through MN4 is as shown in Eq. (8) when W/L of MN4 transistor is M times larger than that of MN3 transistor and the two transistors are all in saturation region. As shown in Eq. (8), the output current $I_O$ of the current mirror circuit mirrors the $I_{REF}$ current M times. If M = 1 then $I_O = I_{REF}$.

$$I_O = \frac{1}{2}\mu_n C_{ox}\frac{M \cdot W}{L}(V_{BIAS} - V_{TN})^2 = M \cdot I_{REF} \tag{8}$$

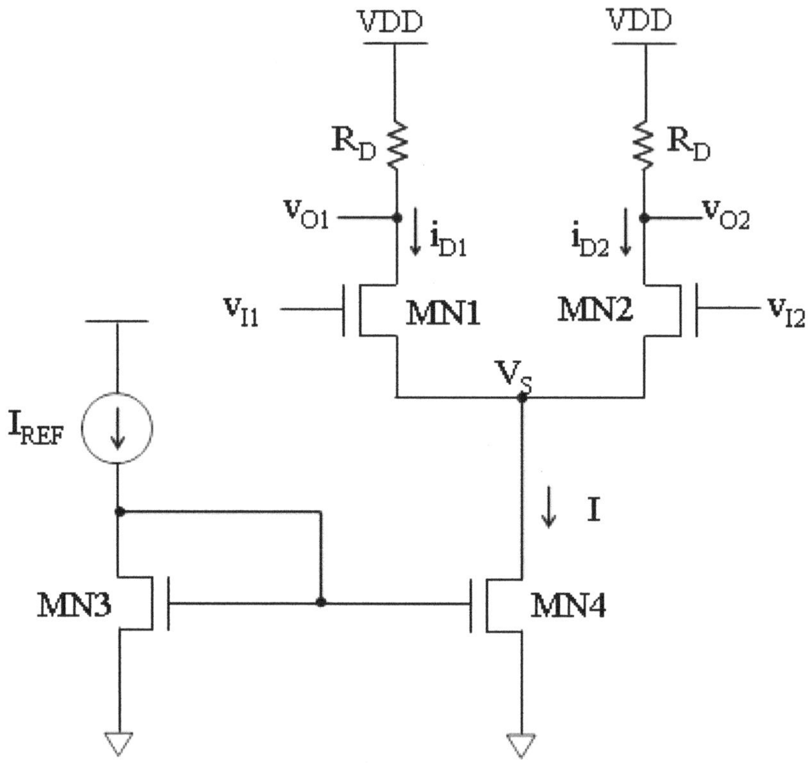


Fig. 4. NMOS differential amplifier circuit using NMOS current mirror as current source.

Fig. 4 shows the case where the NMOS current mirror circuit of Fig. 3 is used instead of the ideal current source in the differential amplifier circuit. In the differential amplifier circuit of Fig. 4, the minimum $V_{CM}$ voltage is the sum of $V_{DSAT.MN4}$, which is the voltage for MN4 to be in the saturation region, and $V_{TN.MN1} + V_{DSAT.MN1}$, which is the voltage for MN1 transistor to be in the saturation region, based on VSS (=0V) as shown in Eq. (9). $V_{TN.MN1}$ is the threshold voltage of the MN1 transistor, and $V_{DSAT.MN1}$ and $V_{DSAT.MN4}$ are the $V_{DSAT}$ voltages, which are the over-driving voltages of MN1 and MN4, respectively. The common mode input voltage range for the differential amplifier is shown in Eq. (9).

$$V_{DSAT.MN4} + V_{TN.MN1} + V_{DSAT.MN1} \leq V_{CM} \leq VDD - \frac{I}{2}R_D + V_{TN.MN1} \qquad (9)$$

In the differential amplifier circuit of Fig. 4, the maximum $V_{CM}$ voltage is the voltage for the MN1 transistor to be in the saturation region based on $VDD$. Since the drain voltage of MN1 transistor for the common mode input voltage $V_{CM}$ is $VDD$ - $R_D$ · (I/2), the maximum voltage of the $V_{CM}$ for the MN1 transistor to be in the saturation region is $VDD$ - $R_D$ · (I/2) + $V_{TN.MN1}$.

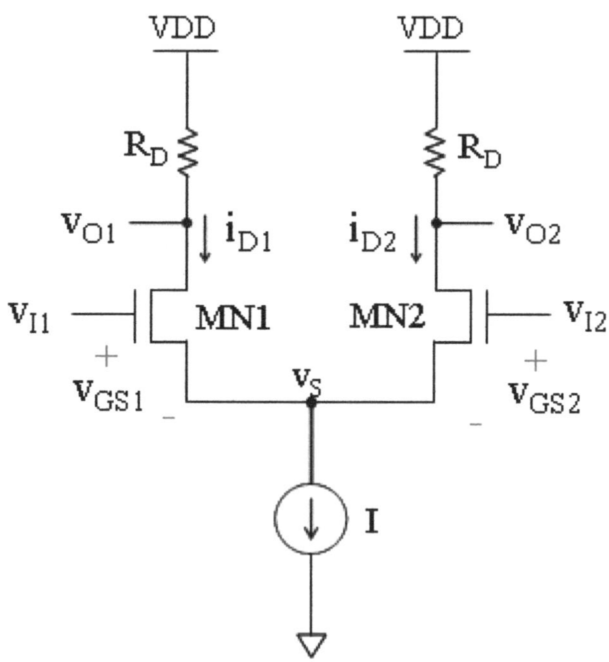


Fig. 5. In case that a differential mode input voltage is applied to a differential amplifier having resistive load as load.

As shown in Fig. 5, assume that $v_{I1}$ and $v_{I2}$ are applied to the inputs of the NMOS differential pair. Since the differential input voltage is $v_{id}=v_{I1}-v_{I2}$, the voltages of $v_{GS1}$ and $v_{GS2}$ are $V_{GS}+v_{id}/2$ and $V_{GS}-v_{id}/2$, respectively. If the vid voltage is a positive voltage, $v_{GS1}$ is higher than $v_{GS2}$ and $i_{D1}$ is larger than $i_{D2}$, so that the $v_{O1}$ voltage is lower than the $v_{O2}$ voltage. Since the differential output voltage $v_{od}$ is $v_{O1}-v_{O2}$, a negative voltage is outputted. If the $v_{id}$ voltage is a negative voltage, the $v_{O1}$ voltage is higher than the $v_{O2}$ voltage because $v_{GS1}$ is lower than $v_{GS2}$ and $i_{D1}$ is less than $i_{D2}$. Thus, the differential output voltage $v_{od}$ outputs a positive voltage. Thus, the differential amplifier circuit of Fig. 5 outputs a differential output voltage $v_{od}$ proportional to the differential input voltage $v_{id}$ to the differential output pair when the differential mode input signal is applied, and has a negative voltage gain.

In the circuit shown in Fig. 1, MN1 and MN2 of the differential input pair are identical. When there is no channel length modulation effect and body effect, when the large signal voltages $v_{I1}$ and $v_{I2}$ are applied, then $i_{D1}$ and $i_{D2}$ are derived as a function of $v_{id}$. Taking roots in Eqs. (10) and (11) and subtracting them, we can obtain Eq. (12).

In the circuit shown in Fig. 1, MN1 and MN2 of the differential input pair are identical. When there is no channel length modulation effect and body effect, when the large signal voltages $v_{I1}$ and $v_{I2}$ are applied, then $i_{D1}$ and $i_{D2}$ are derived as a function of $v_{id}$. Taking roots in Eqs. (10) and (11) and subtracting them, we can obtain Eq. (12).

$$i_{D1}=\frac{1}{2}\mu_n C_{ox}\frac{W}{L}(v_{GS1}-V_{TN})^2 \tag{10}$$

$$i_{D2}=\frac{1}{2}\mu_n C_{ox}\frac{W}{L}(v_{GS2}-V_{TN})^2 \tag{11}$$

$$\sqrt{i_{D1}} - \sqrt{i_{D2}} = \sqrt{\frac{1}{2}\mu_n C_{ox}\frac{W}{L}}(v_{I1} - v_{I2}) \tag{12}$$

Since $v_{I1} - v_{I2}$ is $v_{id}$, Eq. (12) can be written as Eq. (13). In Eq, (13), K is given by Eq. (14).

$$\sqrt{i_{D1}} - \sqrt{i_{D2}} = \sqrt{\frac{1}{2}\mu_n C_{ox}\frac{W}{L}}\, v_{id} = \sqrt{K}\, v_{id} \tag{13}$$

$$K = \frac{1}{2}\mu_n C_{ox}\frac{W}{L} \tag{14}$$

On the other hand, if the KCL (Kirchhoff's Current Law) is applied at the source node of MN1 and MN2, the sum of $i_{D1}$ and $i_{D2}$ becomes I as shown in Eq. (15). The solutions of $i_{D1}$ and $i_{D2}$ are obtained from Eqs. (13) and (15) as Eqs. (16) and (17).

$$i_{D1} + i_{D2} = I \tag{15}$$

$$i_{D1} = \frac{I}{2} + \sqrt{2KI}\left(\frac{v_{id}}{2}\right)\sqrt{1 - \frac{(v_{id}/2)^2}{(I/2K)}} \tag{16}$$

$$i_{D2} = \frac{I}{2} - \sqrt{2KI}\left(\frac{v_{id}}{2}\right)\sqrt{1 - \frac{(v_{id}/2)^2}{(I/2K)}} \tag{17}$$

The drain current equation of the MOS transistor at the DC operating point is given by Eq. (18). Eqs. (19) and (20) can be rewritten from Eqs. (16) and (17) by using Eq. (18).

$$\frac{I}{2} = K(V_{GS} - V_{TN})^2 \tag{18}$$

$$i_{D1} = \frac{I}{2} + \left(\frac{I}{V_{GS} - V_{TN}}\right)\left(\frac{v_{id}}{2}\right)\sqrt{1 - \left(\frac{v_{id}/2}{V_{GS} - V_{TN}}\right)^2} \tag{19}$$

$$i_{D2} = \frac{I}{2} - \left(\frac{I}{V_{GS} - V_{TN}}\right)\left(\frac{v_{id}}{2}\right)\sqrt{1 - \left(\frac{v_{id}/2}{V_{GS} - V_{TN}}\right)^2} \tag{20}$$

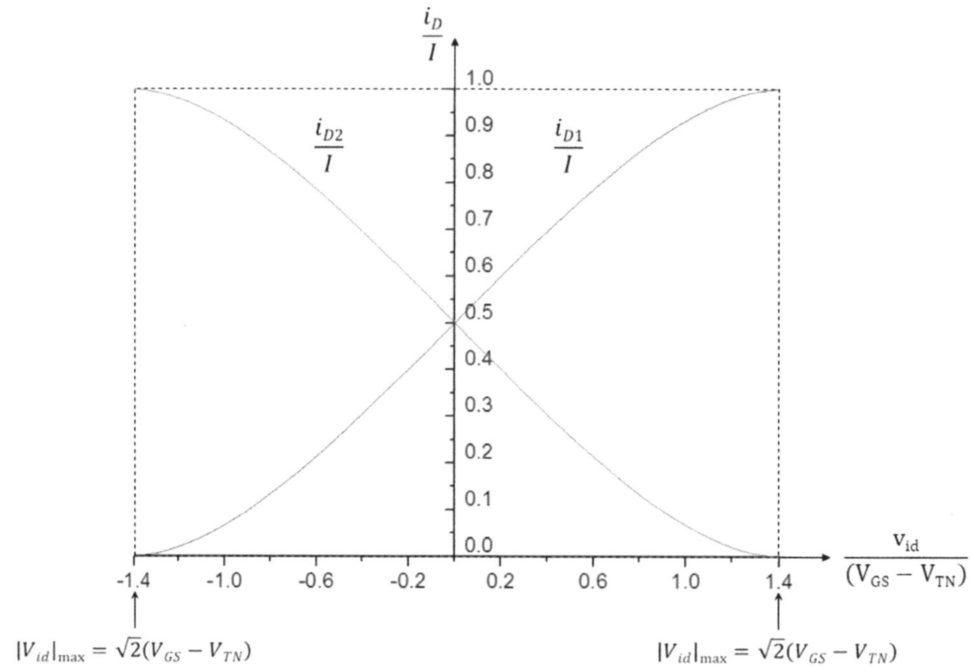


Fig. 6. The normalized current $i_{D1}/I$ and $i_{D2}/I$ curve for the $v_{id}/(V_{GS} - V_{TN})$ voltage.

Fig. 6 shows the curves of the normalized currents $i_{D1}/I$ and $i_{D2}/I$ for the $v_{id}/(V_{GS}-V_{TN})$ voltage using Eqs. (19) and (20). In Fig. 6, the full-swing $v_{id}$ voltage occurs when $i_{D1}=I$ and $i_{D2}=0$ or $i_{D1}=0$ and $i_{D2}=I$. Thus, Eq. (22) satisfying Eq. (21) is derived.

$$\left(\frac{I}{V_{GS}-V_{TN}}\right)\left(\frac{v_{id}}{2}\right)\sqrt{1-\left(\frac{v_{id}/2}{V_{GS}-V_{TN}}\right)^2}=\frac{I}{2} \tag{21}$$

$$|v_{id}|_{\max}=\sqrt{2}(V_{GS}-V_{TN}) \tag{22}$$

Eqs. (19) and (20) can be approximated by Eqs (23) and (24) when the $v_{id}$ voltage is much smaller than $V_{GS}-V_{TN}$, which is a small-signal approximation condition.

$$i_{D1}\cong\frac{I}{2}+\left(\frac{I}{V_{GS}-V_{TN}}\right)\left(\frac{v_{id}}{2}\right)=\frac{I}{2}+g_m\left(\frac{v_{id}}{2}\right) \tag{23}$$

$$i_{D2}\cong\frac{I}{2}-\left(\frac{I}{V_{GS}-V_{TN}}\right)\left(\frac{v_{id}}{2}\right)=\frac{I}{2}-g_m\left(\frac{v_{id}}{2}\right) \tag{24}$$

The low-frequency small-signal equivalent circuit using the hybrid-$\pi$ model for the differential amplifier circuit of Fig. 4 is shown in Fig. 7. Since the small-signal equivalent circuit in Fig. 7 is a linear circuit, the principle of superposition is applied. Therefore, the differential mode voltage gain $A_{vd}$ and the common mode voltage gain $A_{vc}$ are obtained, and then the output voltages $v_{o1}$ and $v_{o2}$ can be calculated as Eqs. (25) and (26) by applying the superposition principle. In Eqs (25) and (26), $v_{id}$ and $v_{ic}$ are given by Eqs. (27) and (28), respectively.

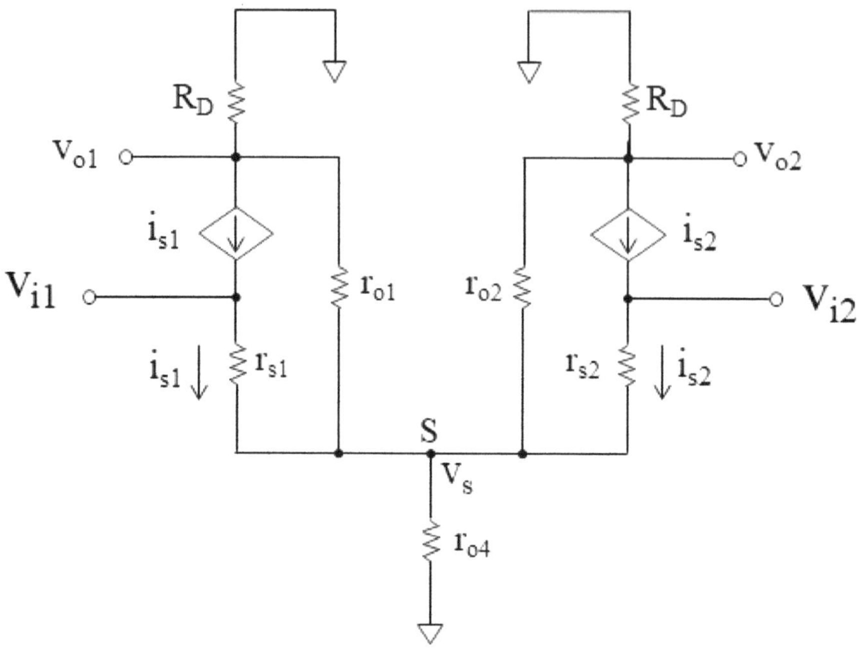


Fig. 7. Low-frequency small-signal equivalent circuit of differential amplifier circuit with resistive load as load.

$$v_{o1} = \frac{1}{2} A_{vd} \cdot v_{id} + A_{vc} \cdot v_{ic} \tag{25}$$

$$v_{o2} = -\frac{1}{2} A_{vd} \cdot v_{id} + A_{vc} \cdot v_{ic} \tag{26}$$

$$v_{id} = v_{i1} - v_{i2} \tag{27}$$

$$v_{ic} = \frac{v_{i1} + v_{i2}}{2} \tag{28}$$

Eqs. (29) and (30) can be obtained by applying the KCL to $v_{o1}$ and $v_{o2}$, respectively.

$$-\left(\frac{1}{R_D}+\frac{1}{r_{o1}}\right)\bullet (v_{o1}+v_{o2})+\left(\frac{2}{r_{s1}}+\frac{2}{r_{o1}}\right)\bullet v_s=0 \quad (29)$$

$$-\frac{1}{r_{o1}}\bullet (v_{o1}+v_{o2})+\left(\frac{2}{r_{s1}}+\frac{2}{r_{o1}}+\frac{2}{r_{o4}}\right)\bullet v_s=0 \quad (30)$$

From Eqs. (29) and (30), we can obtain a relation of $v_s=0$ and $v_{o1}+v_{o2}=0$. If the differential input voltage ($v_{i1}=v_{id}/2$ and $v_{i2}=-v_{id}/2$) is applied, the small-signal voltage at point S becomes 0 and $v_{o1}$ and $v_{o2}$ are equal in magnitude and out of phase with each other as in Eq. (31).

$$v_{o1}=-v_{o2}=\frac{v_{od}}{2} \quad (31)$$

As shown in Fig. 7, when differential mode input voltage $\pm v_{id}/2$ is applied and $v_s=0$ and $v_{o1}=-v_{o2}$ are used, the differential mode ac equivalent circuit of the differential amplifier circuit having resistance load can be drawn as shown in Fig. 8.

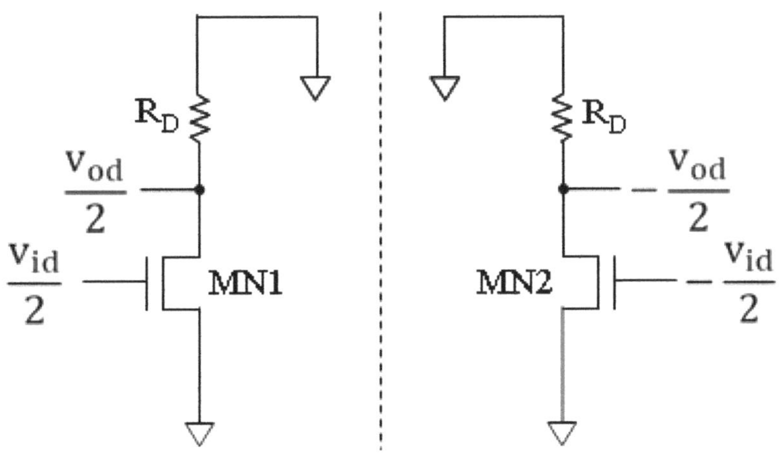


Fig. 8. Differential mode ac equivalent circuit of a differential amplifier circuit with resistive load.

Fig. 9(a) is a differential mode half-circuit made up of the left half of the differential mode ac equivalent circuit of Fig. 8. Fig. 9(b) shows a low frequency small-signal equivalent circuit using the hybrid-$\pi$ model for the differential mode half-circuit of Fig. 9(a).

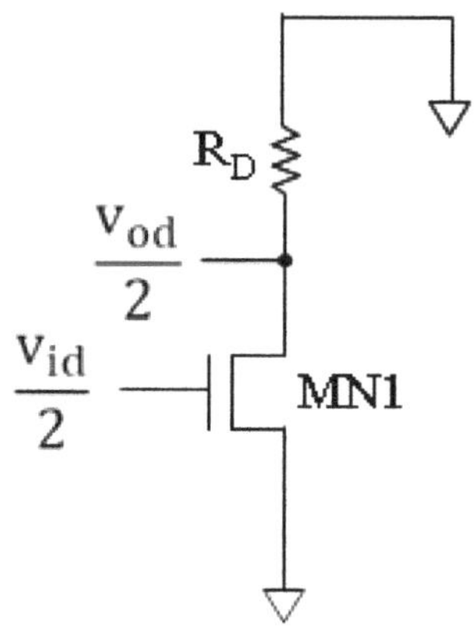


(a) differential mode half-circuit

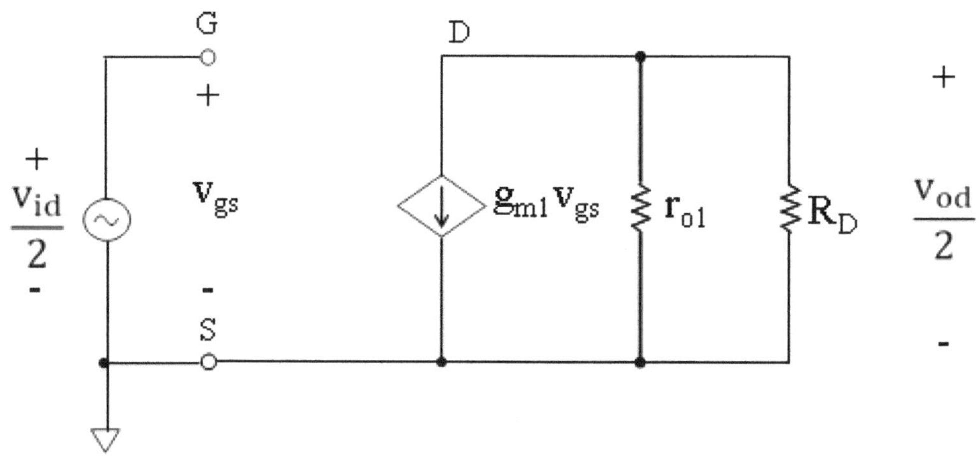


(b) low frequency small-signal equivalent circuit using hybrid-π model

Fig. 9. The differential mode ac equivalent circuit analysis of Fig. 8.

Applying the KVL to the input loop results in Eq. (32).

$$v_{gs} = \frac{v_{id}}{2} \tag{32}$$

Applying the KCL to the drain node of Fig. 9(b) yields Eq. (33).

$$g_{m1}v_{gs} + \frac{v_{od}/2}{r_{o1}} + \frac{v_{od}/2}{R_D} = 0 \tag{33}$$

Substituting Eq. (32) into Eq. (33) yields Eq. (34).

$$g_{m1}v_{id} = \left(\frac{1}{r_{o1}} + \frac{1}{R_D}\right)v_{od} = \frac{v_{od}}{r_{o1}||R_D} \tag{34}$$

If $r_{o4}$ is much larger than $R_D$, $A_{vd}$ can be approximated as Eq. (35).

$$A_{vd} = \frac{v_{od}}{v_{id}} = -g_{m1}(r_{o1}||R_D) \cong -g_{m1}R_D \tag{35}$$

Fig. 10 shows the common mode differential amplifier circuit in which the voltages $v_{i1}$ and $v_{i2}$ are applied to the $v_{ic}$ voltage. When common mode input voltage is applied to the $v_{i1}$ and $v_{i2}$ respectively, the common mode ac equivalent circuit of the differential amplifier circuit having resistance load can be drawn as shown in Fig. 10. A common mode half-circuit made of the left half of the common mode ac equivalent circuit can be made in a similar way to the differential mode.

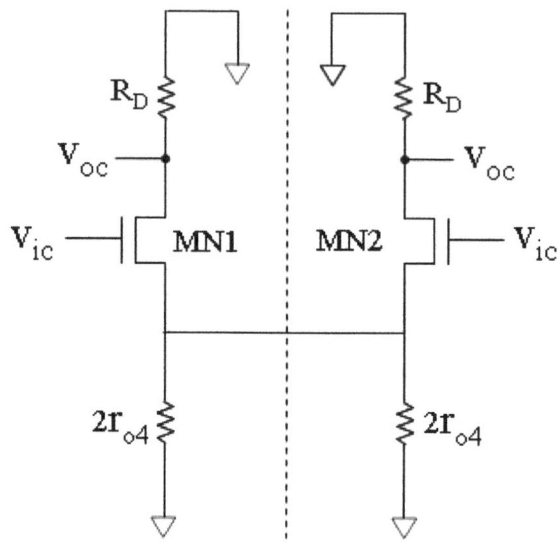


Fig. 10. Common mode equivalent circuit of a differential amplifier circuit with resistive load as load.

Fig. 11(a) is a common mode half-circuit made from the left half of the common mode ac equivalent circuit shown in Fig. 10. Fig. 11(b) shows a low-frequency small-signal equivalent circuit using the hybrid-$\pi$ model for the common mode half-circuit of Fig. 11(a).

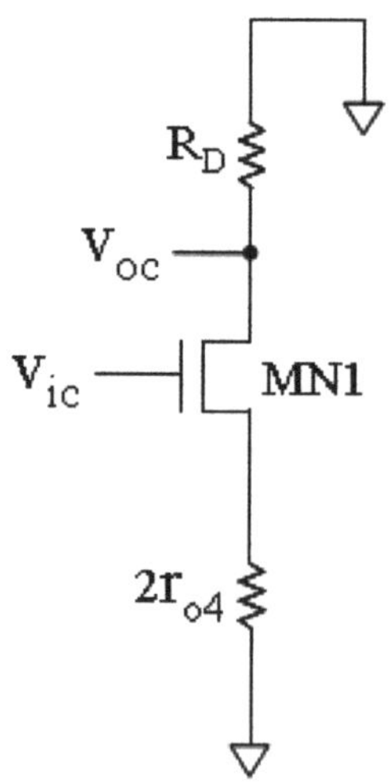


(a) common mode half-circuit

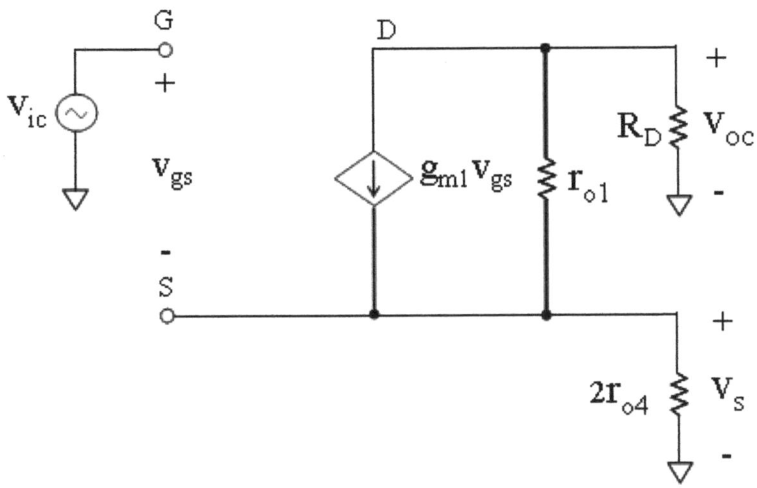


(b) low frequency small-signal equivalent circuit using hybrid-π model

Fig. 11. Common mode ac equivalent circuit analysis of Fig. 10.

Applying the KVL to the input loop of Fig. 11(b) results in Eq. (36).

$$-v_{ic} + v_{gs} + v_s = 0 \tag{36}$$

Applying the KCL to the source and drain nodes yields Eqs. (37) and (38), respectively.

$$-g_{m1}v_{gs} + \frac{v_s - v_{oc}}{r_{o1}} + \frac{v_s}{2r_{o4}} = 0 \tag{37}$$

$$g_{m1}v_{gs} + \frac{v_{oc} - v_s}{r_{o1}} + \frac{v_{oc}}{R_D} = 0 \tag{38}$$

Eqs. (37) and (38) are taken together to obtain Eq. (39).

$$\frac{v_s}{2r_{o4}} = -\frac{v_{oc}}{R_D} \tag{39}$$

Eq. (39) is rewritten as Eq. (40).

$$v_s = -\frac{2r_{o4}}{R_D} v_{oc} \tag{40}$$

If $v_{gs}$ is rewritten in Eq. (36), Eq. (41) is obtained.

$$v_{gs} = v_{ic} - v_s = v_{ic} + \frac{2r_{o4}}{R_D} v_{oc} \tag{41}$$

Substituting Eqs. (40) and (41) into Eq. (37) gives Eq. (42), which can be summarized as Eq. (43).

$$-g_{m1} v_{ic} - \frac{2g_{m1} r_{o4}}{R_D} v_{oc} - \frac{2r_{o4}}{r_{o1} R_D} v_{oc} - \frac{1}{r_{o1}} v_{oc} - \frac{1}{R_D} v_{oc} = 0 \tag{42}$$

$$g_{m1} v_{ic} = -\frac{r_{o1} + R_D + 2r_{o4} + 2g_{m1} r_{o1} r_{o4}}{r_{o1} R_D} v_{oc} \tag{43}$$

On the other hand, the small-signal common mode voltage gain $A_{vc}$ can be written as Eq. (44). Since $2g_{m1} r_{o1} r_{o4}$ is much larger than $r_{o1}$, $r_{o4}$, and $R_D$ in the denominator of Eq. (44), it can be approximated as in Eq. (45).

$$A_{vc} = \frac{v_{oc}}{v_{ic}} = -\frac{g_{m1} r_{o1} R_D}{r_{o1} + R_D + 2r_{o4} + 2g_{m1} r_{o1} r_{o4}} \tag{44}$$

$$A_{vc} = \frac{v_{oc}}{v_{ic}} \cong -\frac{R_D}{2r_{o4}} \tag{45}$$

Fig. 12(b) shows the low-frequency small-signal equivalent circuit using the T-equivalent model of Fig. 12(a) in the common mode half-circuit of Fig. 12(a). The T-equivalent model in Fig. 12(a) ignores the small-signal output resistance $r_o$ between drain and source.

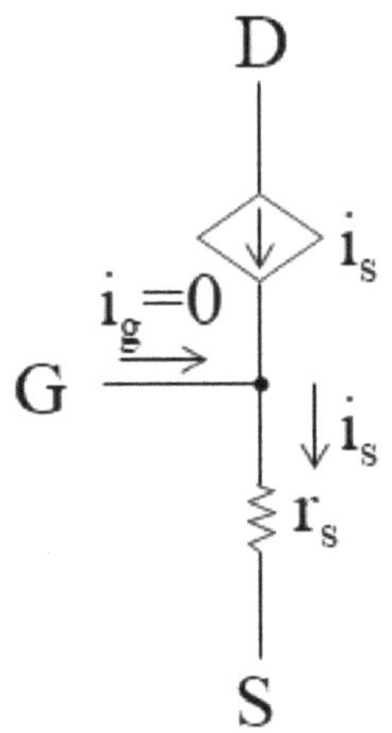


(a) T-equivalent model

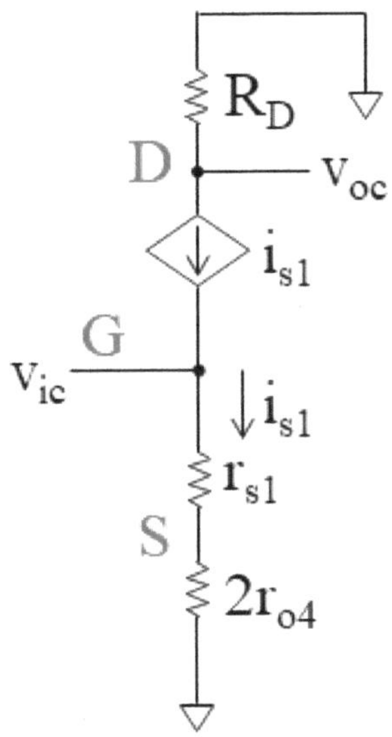


(b) low-frequency small-signal equivalent circuit using T-equivalent model

Fig. 12. Equivalent circuit analysis of common mode ac equivalent circuit with T-equivalent model applied.

In Fig. 12, $i_{s1}$ is equal to Eq. (46) and $v_{oc}$ is equal to Eq. (47). Substituting Eq. (46) into Eq. (47) yields Eq. (48). The small-signal common mode voltage gain $A_{vc}$ can be written as in Eq. (49). In the denominator of Eq. (48), since $2r_{o4}$ is much larger than $r_{s1}$, it can be approximated as Eq. (50).

$$i_{s1} = \frac{v_{ic}}{r_{s1} + 2r_{o4}} \tag{46}$$

$$v_{oc} = -R_D i_{s1} \tag{47}$$

$$v_{oc} = -\frac{R_D}{r_{s1} + 2r_{o4}} v_{ic} \tag{48}$$

$$A_{vc} = \frac{v_{oc}}{v_{ic}} = -\frac{R_D}{r_{s1} + 2r_{o4}} \tag{49}$$

$$A_{vc} \cong -\frac{R_D}{2r_{o4}} \tag{50}$$

As shown in Fig. 11(b), whether the hybrid-$\pi$ model or the T-equivalent model is applied to the NMOS transistor MN1 as shown in Fig. 12(b), the same result for the common mode voltage gain can be obtained as shown in Figs. (45) and (50).

In differential amplifiers, CMRR (Common Mode Rejection Ratio) is a numerical value that indicates how well the differential mode signal is amplified and the common mode signal is rejected. CMRR is the absolute value of the differential mode voltage gain $A_{vd}$ divided by the common mode voltage gain $A_{vc}$ as shown in Eq. (51).

$$CMRR = \left| \frac{A_{vd}}{A_{vc}} \right| \tag{51}$$

The CMRR for the differential amplifier circuit with the resistive load is given by Eq. (52).

$$CMRR = \left| \frac{A_{vd}}{A_{vc}} \right| \cong 2g_{m1}r_{o4} \tag{52}$$

## 6.2 Differential Amplifier with Active Load

Fig. 13 shows a differential amplifier circuit with a PMOS current mirror as an active load. The differential input voltage $v_{I1} - v_{I2}$ is amplified and then the single-ended output voltage $v_O$ is outputted. In the differential amplifier circuit with active load, common mode inputs are canceled each other, and the output voltage $v_O$ is twice as much as the differential amplifier circuit with the resistive load for the differential mode input signal. In addition, small-signal output resistance $R_o$ is large and voltage gain is large. Thus, the differential amplifier circuit with active load has better CMRR characteristic than the differential amplifier circuit with resistive load.

The differential amplifier circuit with the PMOS current mirror in Fig. 13 is not symmetrical because it uses the PMOS current mirror as load instead of the resistor $R_D$. Therefore, the differential amplifier circuit with active load can not apply the half-circuit concept when calculating the voltage gain. In general, the differential amplifier circuit with active load can be easily analyzed by applying $G_m$ and $R_o$.

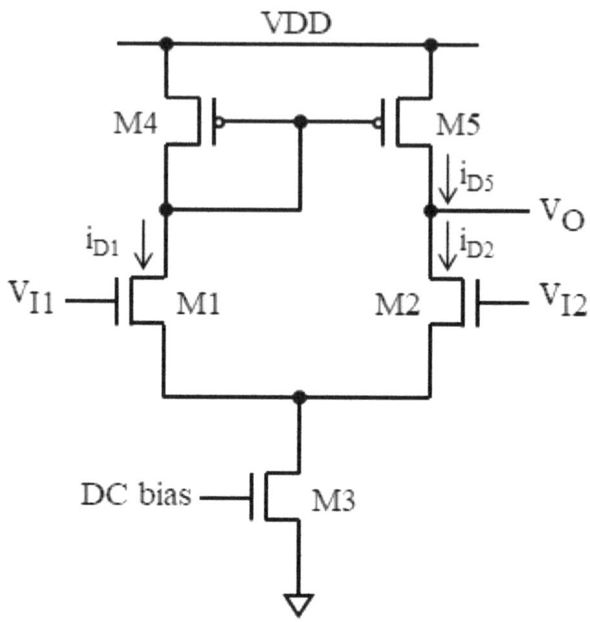


Fig. 13. Differential amplifier circuit with PMOS current mirror as load.

Fig. 14 shows the small-signal equivalent circuit for calculating $R_o$ for the differential amplifier circuit of Fig. 13. The small signal output resistance $R_o$ is the resistance at the output node when $v_{i1} = v_{i2} = 0$.

In Fig. 14, the $V_{SG}$ voltages of M4 and M5 in the form of PMOS current mirrors are the same and are identical transistors, so it is same as Eq. (53).

$$i_{s5} = i_{s4} \tag{53}$$

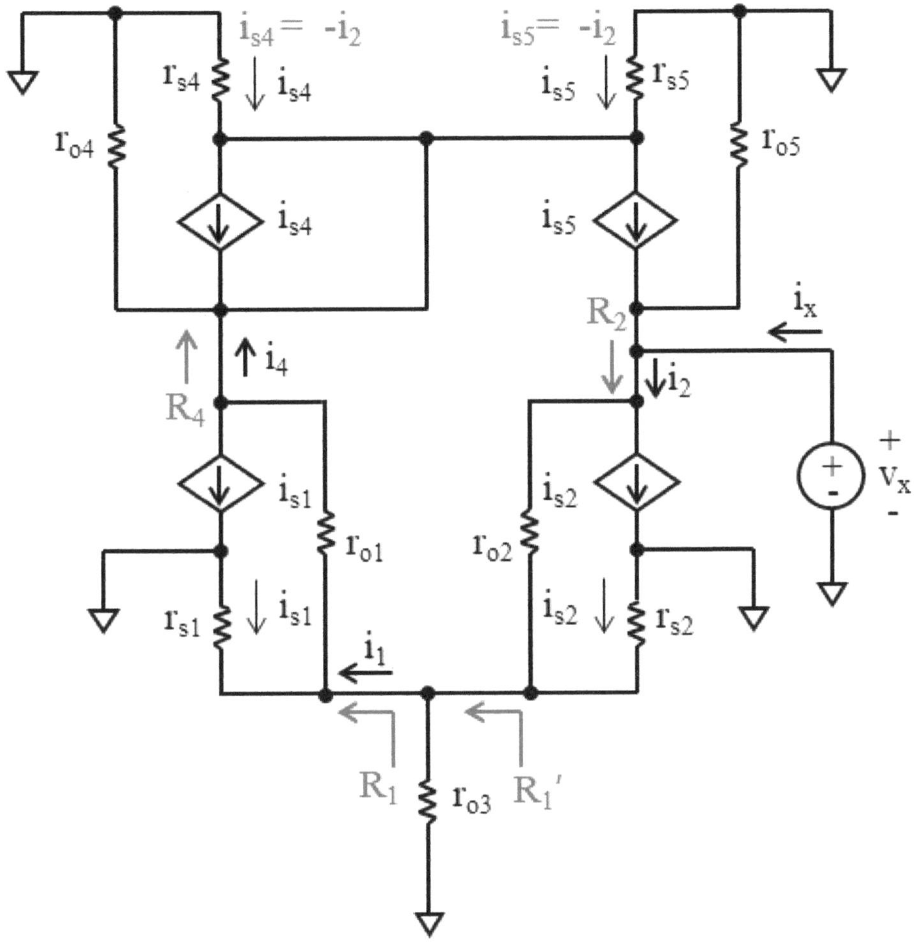


Fig. 14. Small-signal equivalent circuit for calculating $R_o$ for the differential amplifier circuit of Fig. 13.

The small-signal resistance $R_4$ seen from the drain node of the PMOS diode M4 in the PMOS current mirror is as shown in Eq. (54).

$$R_4 = r_{s4} || r_{o4} \cong r_{s4} \tag{54}$$

And in Fig. 14, the small-signal resistance $R_1$ seen from the source node of M1 is same as Eq. (55). Eq. (55) is calculated using the small-signal input resistance $R_i$ of the common gate amplifier.

$$R_1 = \frac{1}{g_{m1}}\left(1 + \frac{r_{s4}}{r_{o1}}\right) \cong \frac{1}{g_{m1}} = r_{s1} \tag{55}$$

$R1'$ is the same as Eq. (56) because $R_1$ and $r_{o3}$ are parallel.

$$R_1' = r_{o3} || R1 = r_{o3} || r_{s1} \cong r_{s1} \tag{56}$$

The resistance $R_2$ seen from the M2 drain node in Fig. 14 can be obtained from the equivalent circuit of Fig. 15.

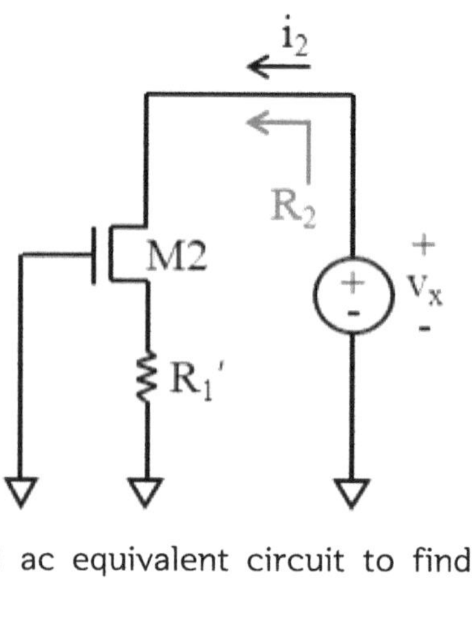


(a) ac equivalent circuit to find $R_2$.

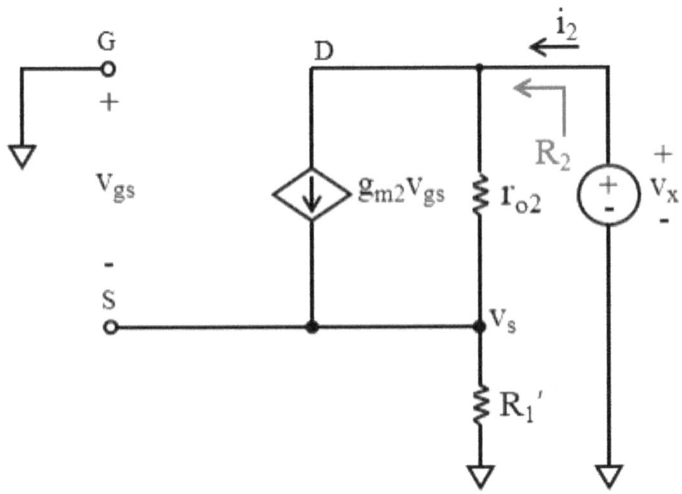


(b) ac equivalent circuit using hybrid-π model to find $R_2$

Fig. 15. Equivalent circuit to find small-signal output resistance $R_2$ of Fig. 14.

The source node voltage $v_s$ of Fig. 15(b) is equal to Eq. (57), and $v_{gs}$ is equal to Eq. (58). If the KCL is applied to the drain node of Fig. 57(b), it is same as Eq. (59). Rearranging Eq. (59) for $i_2$ yields Eq. (60). $R_2$ is calculated using Eq. (60), which is same as Eq. (61).

$$v_s = R_1^{'} i_2 \tag{57}$$

$$v_{gs} = - v_s = - R_1^{'} i_2 \tag{58}$$

$$\begin{aligned} i_2 &= g_{m2} v_{gs} + \frac{v_x - v_s}{r_{o2}} \\ &= - g_{m2} R_1^{'} i_2 + \frac{v_x}{r_{o2}} - \frac{R_1^{'}}{r_{o2}} i_2 \end{aligned} \tag{59}$$

$$\left(1 + g_{m2}R_1^{'} + \frac{R_1^{'}}{r_{o2}}\right)i_2 = \frac{v_x}{r_{o2}} \tag{60}$$

$$R_2 = \frac{v_x}{i_2} = r_{o2} + R_1^{'} + g_{m2}r_{o2}R_1^{'} \tag{61}$$

In Eq. (61), $R_1{'}$ is $r_{s1}$, and since M1 and M2 are identical transistors, $R_2$ is approximated to $2r_{o2}$ as in Eq. (62). Meanwhile, $i_2$ is a value obtained by dividing $v_x$ by $R_2$ as shown in Eq. (63).

$$R_2 = r_{o2} + R_1^{'} + g_{m2}r_{o2}R_1^{'} = 2r_{o2} \tag{62}$$

$$i_2 = \frac{v_x}{R_2} = \frac{v_x}{2r_{o2}} \tag{63}$$

In addition, $R_1{'}$ is almost the same as $r_{s1}$ of the M1 transistor as shown in Eq. (56), so $i_1$ is almost equal to $i_2$ as shown in Eq. (64). $i_{s1}$ is almost equal to $-i_2$ as seen in Eq. (65), and eventually $i_4$ in Fig. 14 becomes $i_2$ as seen in Eq. (66). On the other hand, $i_{s4}$ is almost equal to $-i_4$ as in Eq. (67) and eventually becomes $-i_2$. And since M4 and M5 are PMOS current mirrors, $i_{s5}$ is the same as $i_{s4}$ as shown in Eq. (68) and eventually becomes $-i_2$.

$$i_1 \cong i_2 \tag{64}$$

$$i_{s1} \cong -i_1 = -i_2 \tag{65}$$

$$i_4 \cong i_2 \tag{66}$$

$$i_{s4} \cong -i_4 = -i_2 \tag{67}$$

$$i_{s5} = i_{s4} = -i_2 \tag{68}$$

The $i_x$ current in Fig. 14 for obtaining the small-signal output resistance $R_o$ is shown in Eq. (69). So, the small-signal output resistance $R_o$ comes out as the parallel resistance of $r_{o2}$ and $r_{o5}$ as shown in Eq. (70).

$$i_x = i_2 - i_{s5} + \frac{v_x}{r_{o5}} = i_2 + i_2 + \frac{v_x}{r_{o5}} = \frac{v_x}{r_{o2}} + \frac{v_x}{r_{o5}} \tag{69}$$

$$R_o = r_{o2} \| r_{o5} \tag{70}$$

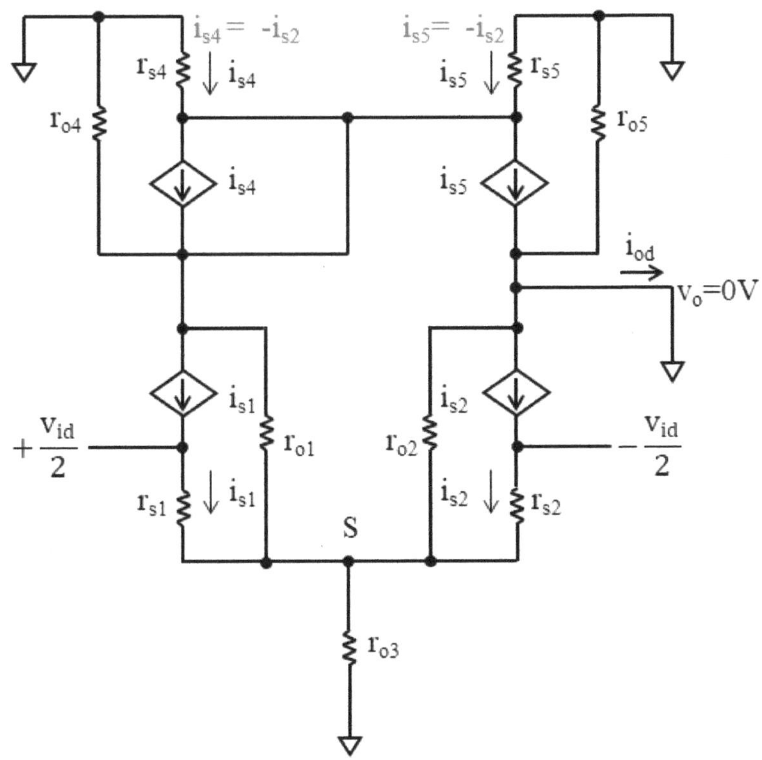


Fig. 16. Small-signal equivalent circuit for calculating common mode transconductance $G_{md}$.

Differential mode transconductance $G_{md}$ calculates the output short circuit current $i_{od}$ for the differential input voltage $v_{id}$. Fig. 16 shows a small-signal equivalent circuit for calculating the differential mode transconductance $G_{md}$. In the small-signal equivalent circuit of Fig. 16, the differential input voltages $v_{i1}$ and $v_{i2}$ are $v_{id}/2$ and $-v_{id}/2$, respectively, and $v_o = 0$. And the source node of MN1 and MN2 is 0V as virtual ground. $i_{s1}$ and $i_{s2}$ in Fig. 16 are given by Eqs. (71) and (72).

$$i_{s1} = \frac{v_{id}/2}{r_{s1}} \tag{71}$$

$$i_{s2} = -\frac{v_{id}/2}{r_{s2}} \tag{72}$$

In addition, since MN1 and MN2 as well as MP4 and MP5 are the identical transistors respectively, $r_{s1} = r_{s2}$ and $r_{s4} = r_{s5}$. The output short circuit current $i_{od}$ is derived as Eq. (73). $G_{md}$ is the differential mode transconductance using Eq. (74).

$$i_{od} = i_{s5} - i_{s2} = i_{s4} - i_{s2} = i_{s1} - i_{s2} = \frac{v_{id}}{r_{s1}} = g_{m1} v_{id} \tag{73}$$

$$G_{md} = \frac{i_{od}}{v_{id}} = g_{m1} \tag{74}$$

Common mode transconductance $G_{mc}$ can be obtained by using the equivalent circuit of Fig. 17 with $v_{i1} = v_{i2} = v_{ic}$, and $v_o = 0\,V$. In Fig. 17, the MOS transistor is replaced by a T-equivalent model that includes $r_o$. The common mode transconductance $G_{mc}$ and CMRR are derived as Eq. (58) and Eq. (59), respectively.

$$G_{mc} = \frac{i_{oc}}{v_{ic}} = -\frac{1}{2 r_{o4} g_{m5} (r_{o1} \| r_{o5})} \tag{75}$$

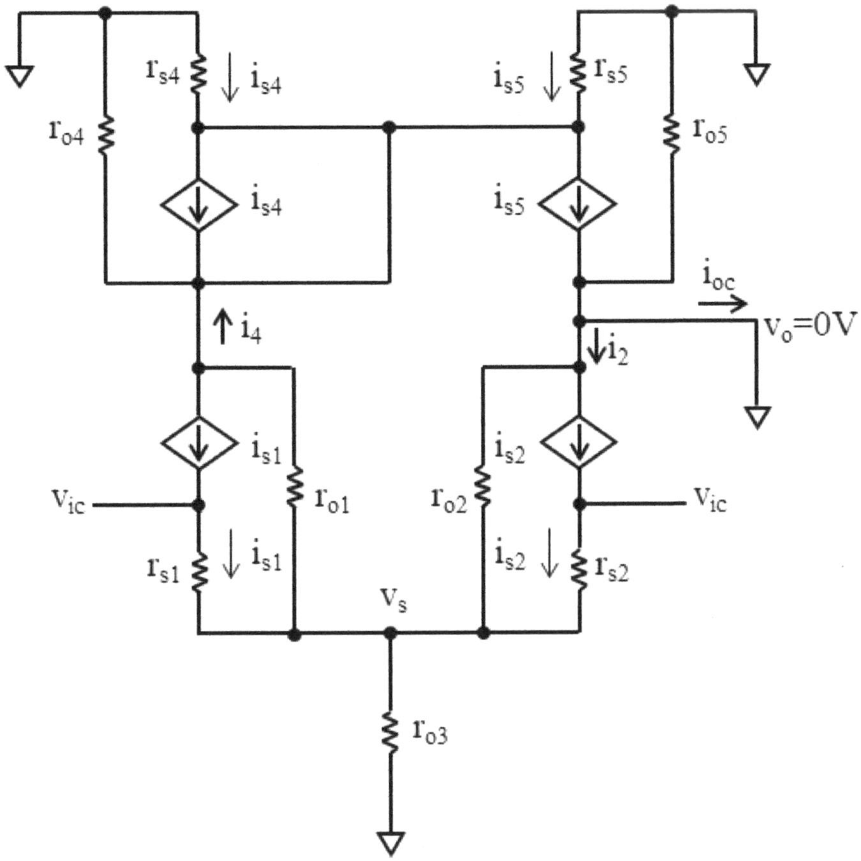


Fig. 17. Small-signal equivalent circuit for calculating common mode transconductance $G_{mc}$.

$$CMRR = \left|\frac{A_{vd}}{A_{vc}}\right| = \left|\frac{G_{md}R_o}{G_{mc}R_o}\right| = \left|\frac{G_{md}}{G_{mc}}\right| = (2g_{m1}r_{o4})\{g_{m5}(r_{o1}||r_{o5})\} \tag{76}$$

**Example 6-1**

*Given that $R_D$=15 kΩ, VDD=5 V, $I_{REF}$=20 μA, $V_{TN}$=0.7 V, $\mu_n C_{ox}$=70 μA/V², and $V_A$=50 V, a fully differential input signal is applied to $v_{I1}$ and $v_{I2}$ in Fig. 18. Answer the following questions.*

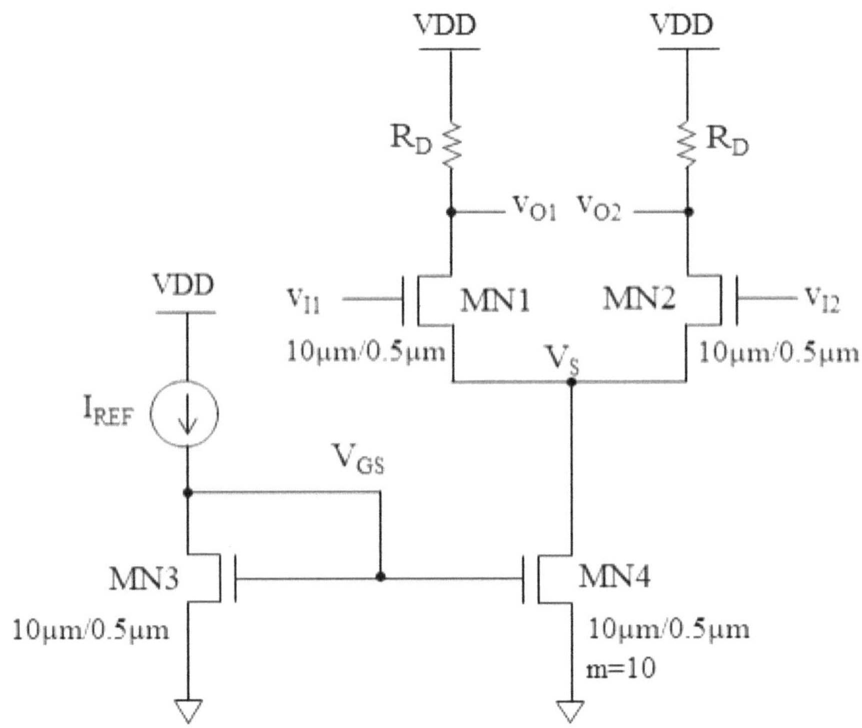


Fig. 18. A differential amplifier for Example 6-1.

(a) Calculate the gate voltage $V_{GS}$ of the NMOS current mirror.

Question (a) is a problem of calculating the $V_{GS}$ voltage of the NMOS current mirror. Since $I_{REF}$ current flows, it can be calculated as shown in Eq. (77).

$$V_{GS} = V_{TN} + \sqrt{\frac{2 \times I_{REF}}{\mu_n C_{ox} \frac{W}{L}}} \tag{77}$$

$$= 0.7\ V + \sqrt{\frac{2 \times 20\ \mu A}{70\ \mu A/V^2 \times \frac{10\ \mu m}{0.5\ \mu m}}}$$

$$= 0.869\ V$$

(b) Calculate $g_{m.MN1}$ and $g_{m.MN4}$. $g_{m.MN1}$ is $g_m$ for the MN1 transistor and $g_{m.MN4}$ is $g_m$ for the MN4 transistor.

Question (b) is a problem of calculating $g_{m.MN1}$ and $g_{m.MN4}$ of MN1 and MN4 transistors. At the DC operating point, the drain currents for MN1 and MN4 are 100 μA and 200 μA, respectively. So $g_{m.MN1}$ and $g_{m.MN4}$ are as shown in Eqs. (78) and (79).

$$g_{m.MN1} = \sqrt{2\mu_n C_{ox} \frac{W}{L} I_{DS \cdot MN1}} \tag{78}$$

$$= \sqrt{2 \times 70\ \mu A/V^2 \times \frac{10\ \mu m}{0.5\ \mu m} \times 100\ \mu A}$$

$$= 0.529\ mS$$

$$g_{m.MN4} = \sqrt{2\mu_n C_{ox} \frac{W}{L} I_{DS \cdot MN4}} \tag{79}$$

$$= \sqrt{2 \times 70\ \mu A/V^2 \times \frac{10 \times 10\ \mu m}{0.5\ \mu m} \times 200\ \mu A}$$

$$= 2.366\ mS$$

**(c) Calculate $r_{o.MN1}$. $r_{o.MN1}$ is $r_o$ for the MN1 transistor.**

Question (c) is a problem of calculating $r_{o.MN1}$ which is the $r_o$ of the MN1 transistor. Since $V_A$ is 50 V, the channel length modulation parameter $\lambda$ is shown in Eq. (80), and $r_{o.MN1}$ is shown in Eq. (81).

$$\lambda = \frac{1}{V_A} = \frac{1}{50\ V} = 0.02\,V^{-1} \tag{80}$$

$$\begin{aligned} r_{o.MN1} &= \frac{1}{\lambda I_{D.MN1}} \\ &= \frac{1}{0.02\,V^{-1} \times 100\ \mu A} \\ &= 500\ K\Omega \end{aligned} \tag{81}$$

**(d) Draw a differential mode half-circuit.**

Question (d) is the problem of drawing a differential mode half circuit. In differential mode, the $V_S$ node of the differential amplifier circuit is a virtual ground node, so it corresponds to ac GND. The half circuit is shown in Fig. 19. The input voltage is $v_{id}/2$ and the output voltage is $v_{od}/2$.

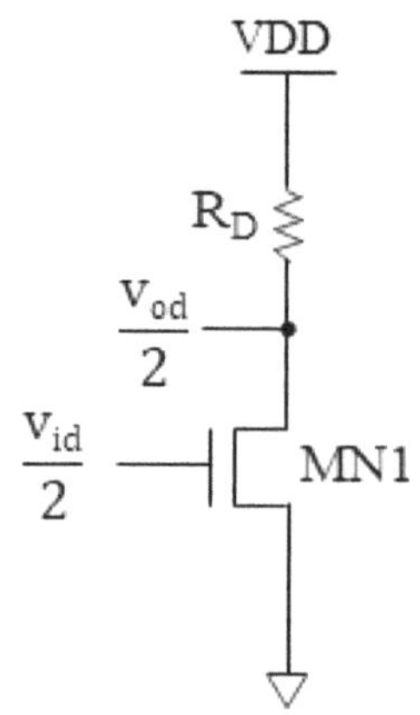


Fig. 19. Differential mode half circuit.

**(e) For the differential mode half-circuit, draw a low-frequency small-signal equivalent circuit based on the hybrid-$\pi$ model.**

Question (e) is a problem of drawing a low-frequency small-signal equivalent circuit by applying a hybrid $\pi$-model to a differential mode half circuit. As shown in Fig. 20, short the DC voltage source VDD in the differential mode half circuit and replace the NMOS transistor with a hybrid $\pi$-model.

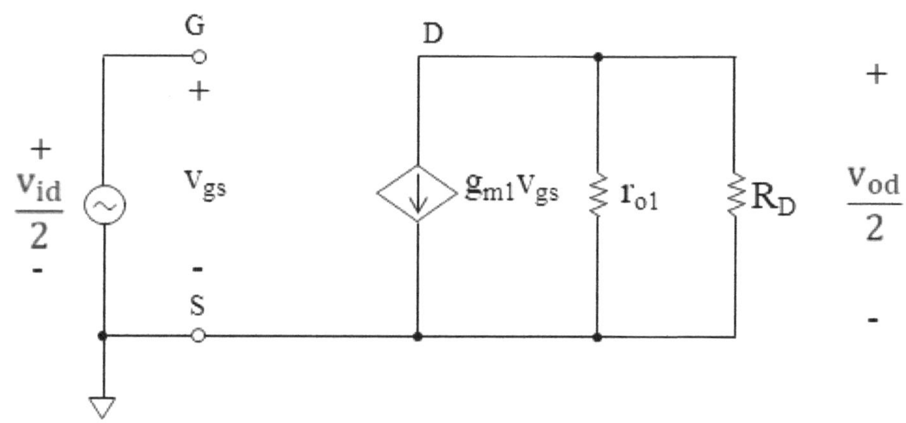


Fig. 20. Low-frequency small-signal equivalent circuit based on the hybrid-π model.

**(f) Using $g_{m.MN1}$ and $r_{o.MN1}$, calculate the differential mode voltage gain $A_{vd}$.**

Question (f) is a problem of calculating the differential mode voltage gain $A_{vd}$ using $g_{m.MN1}$ and $r_{o.MN1}$. Calculating $A_{vd}$ using $g_{m.MN1}$ and $r_{o.MN1}$ is shown in Eq. (65).

$$\begin{aligned} A_{vd} &= \frac{v_{od}}{v_{id}} \\ &= -g_{m.MN1}(r_{o1.MN1} \| R_D) \\ &= -0.529mS \times \frac{500K\Omega \times 15K\Omega}{500K\Omega + 15K\Omega} \\ &= -7.704 \end{aligned} \tag{82}$$

## Exercises

1. Answer the following questions.

   (a) Draw a PMOS differential amplifier circuit using an ideal current source.

   (b) Describe four characteristics of OP AMP.

   (c) Draw an NMOS differential amplifier circuit with a PMOS current mirror as the load using an ideal current source I.

2. Given that $R_D$=30 KΩ, VDD=5 V, $I_{REF}$=10 μA, $V_{TN}$=0.7 V, $\mu_n C_{ox}$=70 μA/V$^2$, and $V_A$ = 50 V, fully differential input signals are applied to $v_{I1}$ and $v_{I2}$.

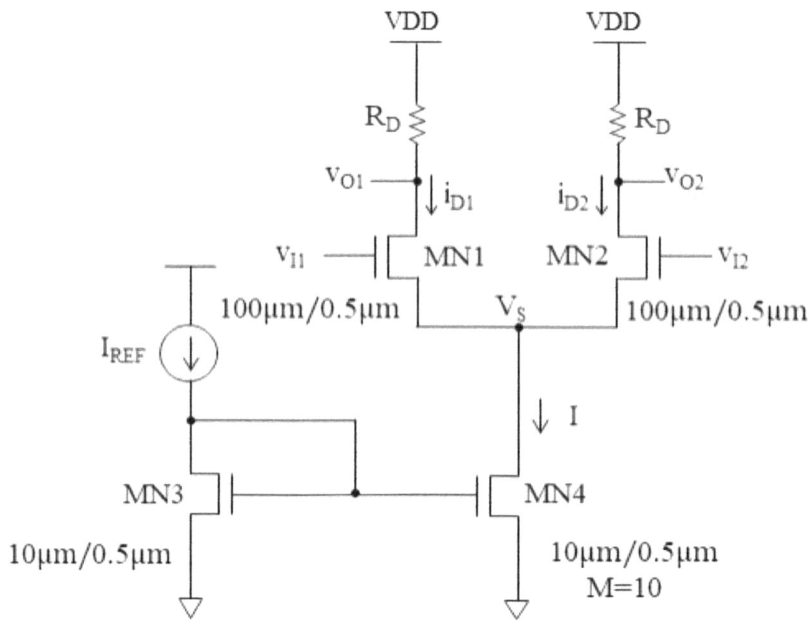

(a) Calculate the gate voltage $V_{GS}$ of the NMOS current mirror.

(b) Calculate $g_{m.MN1}$ and $g_{m.MN4}$. $g_{m.MN1}$ means $g_m$ of MN1 transistor, and $g_{m.MN4}$ means $g_m$ of MN4 transistor.

(c) Calculate $r_{o.MN1}$. $r_{o.MN1}$ means $r_o$ of MN1 transistor.

(d) Draw differential mode half-circuit.

(e) Draw a low-frequency small-signal equivalent circuit when hybrid-$\pi$ model is applied to differential mode half-circuit.

(f) Using $g_{m.MN1}$ and $r_{o.MN1}$, calculate the differential mode voltage gain $A_{vd}$.

3. Given that VDD = 5 V, I = 200 μA, $\mu_p C_{ox}$ = 35 μA/V$^2$, $V_{A.NMOS}$=50 V, and $V_{A.PMOS}$=50 V, a fully differential input to $v_{I1}$, and $v_{I2}$ are applied.

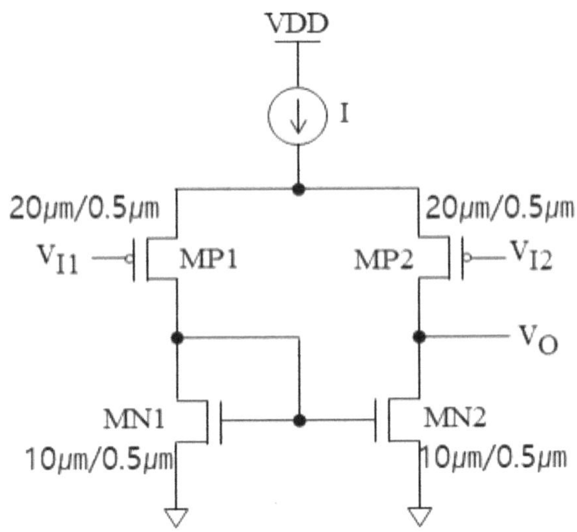

(a) Calculate $g_{m.MN2}$ and $g_{m.MP2}$ . $g_{m.MN2}$ is $g_m$ for the MN2 transistor and $g_{m.MP2}$ is $g_m$ for the MP2 transistor.

(b) Calculate $r_{o.MN2}$. and $r_{o.MP2}$.

(c) Calculate the output resistance $R_o$.

(d) Calculate the differential mode voltage gain $A_{vd}$.

4. Given that $R_D$=15 KΩ, $VDD$=5 V, $I$=200 μA, $V_{TN}$=0.7V, $\mu_n C_{ox}$=70 μA/V², and $V_A$ = 50 V, fully differential input signals are applied to $v_{I1}$ and $v_{I2}$.

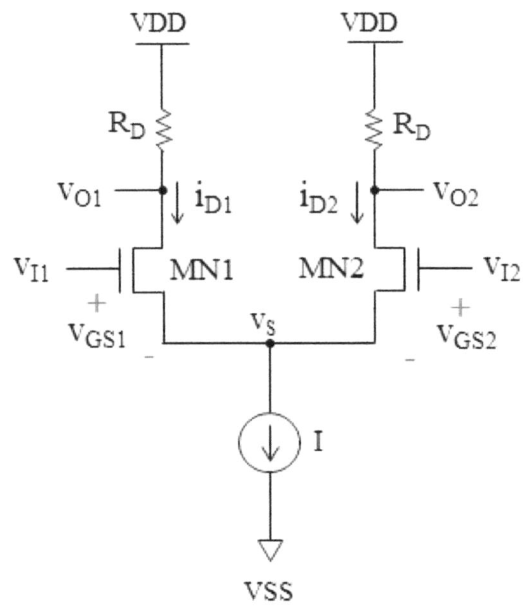

(a) Find the transconductance $g_{m.MN2}$ of the MN2 transistor.

(b) Find $r_{o.MN2}$, the $r_o$ of the MN2 transistor.

(c) Find the differential voltage gain $A_{vd}$ using $g_{m.MN1}$ and $r_{o.MN2}$.

## Chap. 6 References

[1] Hongjun Park, Design of CMOS Analog Integrated Circuits, $\Sigma$Sigma Press, 1999.

[2] B. Razavi, Design of Analog CMOS Integrated Circuits, Mc-Graw Hill Education, 2017.

[3] Sedra et al., Microelectronic Circuits, Oxford.

**YoungHee Kim**

Professor at the Department of Electronic Engineering, Changwon National University

**PanBong Ha**

Professor at the Department of Electronic Engineering, Changwon National University

# CMOS Analog Circuit Design

2020년 9월 10일 초판 인쇄

2020년 9월 15일 인쇄 발행

저 자 김영희, 하판봉

발 행 인 송기수

발 행 처 도서출판 GS인터비전

편 집 처 도서출판 GS인터비전

편 집 인 공예서

표지디자인 김명신

인 쇄 처 트윈벨미디어

등 록 번 호 제 25100-2016-000050호

ISBN 979-11-5576-344-5(93560)

주 소 서울 은평구 증산로 15길 69 2층

전 화 02-976-7898, 02-3272-7898.

팩 스 02-6468-7898

홈페이지 gsintervision.co.kr

E-Mail gsinter7@gmail.com

정 가 20,000원